MOONS OF THE SOLAR SYSTEM

By
Thomas Wm. Hamilton

Strategic Book Publishing and Rights Co.

Strategic Book Publishing and Rights Co.
12620 FM 1960, Suite A4-507
Houston, TX 77065
www.sbpra.com

ISBN: 978-1-62516-175-8

Book Design: Suzanne Kelly

*If any moon discussed below has another moon mentioned
in its write up, the second moon belongs to the same planet unless
a planet name appears in parenthesis after the moon's name.*

MERCURY

Mercury, the planet nearest the Sun, has no moons. It was probably Isaac Asimov (1920-1992) who first pointed out that Mercury cannot have a moon. Any moon more than 3000 miles from the planet would eventually be dragged away by the Sun's gravity. However, Mercury's Roche Limit extends to 3600 miles from the planet. The Roche Limit, discovered by Edouard A. Roche (1820-1883), states that any satellite having the same density as its primary will be pulled apart by the primary if it orbits within 2.44 radii of the primary. Those with a density less than the primary must orbit even further away. So a Mercurian moon would either be pulled away by the Sun or get torn to shreds by the planet (or have an improbably high density). We might note that artificial satellites, such as the MESSENGER spacecraft in orbit around Mercury, are too small for the shredding fate to apply. They have to be about ten miles across before the effect of the planet's different gravitational pull of opposite sides will result in destruction.

VENUS

Venus also has no moons, but lacks Mercury's excuse, as there is room between the Roche Limit and the point where the Sun would drag away a satellite for one to exist. The reason for the lack may be that no moon formed in the fairly small region allowed.

EARTH

[T 1] (The codes in brackets are used for the various indexes. Each planet has a letter, with a number for each moon.) Earth has one Moon. It is the only moon in the Solar System that did not have to be discovered. Dinosaurs could see it, assuming they stayed up at night, or looked in the appropriate part of the daytime sky when it was up.

PHASES:

As our Moon goes around Earth, it does not make its own light, so we are only able to see that portion of the Moon which is both illuminated by the Sun and also facing Earth. When the Moon falls on the straight line Sun—Moon—Earth, we have the night side of the Moon facing us. This is called the New Moon, and is not visible. The New Moon rises with the Sun, and sets with the Sun, another good reason for it not to be visible.

As the Moon moves in its orbit, we gradually begin to see a small portion of the lit side. This phase is known as waxing crescent (the word waxing is related to the German word wachsen, meaning to grow). A waxing crescent rises shortly after sunrise, but is rarely visible until evening, setting shortly after sunset.

Once half the lit up side of the Moon is visible we have First Quarter. Half the Moon is always lit by the Sun; here we are seeing half of the lit side. Half of half is a quarter. The only Half Moon was Henry Hudson's ship, there is *no* lunar phase with that name. First Quarter Moons rise at noon and set at midnight.

Following the quarter phase we get waxing gibbous. Gibbous is sort of a lopsided looking moon, and it is said that artists almost never depict a gibbous moon because it is esthetically displeasing. The waxing gibbous moon rises in the afternoon, and sets between midnight and dawn.

The next phase is Full. Here the Sun—Earth—Moon form a straight line. The Full Moon rises at sunset, and sets at sunrise, so it is the only lunar phase visible all night. The Full Moon reaches a magnitude of −12.9.

The waning gibbous moon rises after sunset but before midnight, and sets sometime during the morning.

Third or Last Quarter Moon rises at midnight and sets at noon. People unaware of how phases work and the relationship with the Sun often see this phase or waning gibbous and are bewildered to learn the Moon is visible in the daytime.

Waning crescent phases rise between midnight and sunrise, and set in the afternoon.

After waning crescent the Moon returns to the New phase. The complete cycle takes an average of 29.53 days to complete. Many different ancient peoples, including the Romans, Jews, Egyptians and Chinese created calendars with the lunar month. Most handled the fractional day by alternating 29 day months with 30 day months, or used some other formula for keeping things fairly even.

ORBIT

The Moon's orbit around the Earth is not a perfect circle. Its closest approach to Earth, known as perigee, is around 220,000 miles, while the most distant, or apogee, is about 252,000 miles. These numbers are slightly different every month, being influenced by the Sun's gravity, and to a very minor extent the gravity of a few of the planets. The Moon's average distance is 238,487 miles.

The Moon's orbit is tilted with respect to both the Earth's equator and to Earth's orbit around the Sun. This is fairly unusual among large, close moons, many of which have their orbits aligned with their planet's equator. In our case the Moon's orbit is tilted 5.3 degrees to the plane of Earth's orbit. Our equator is, of course, tilted 23.5 degrees to our orbit, so the Moon can vary between 18 and 29 degrees from our equator. The main effect of this is on eclipses, which are discussed below.

The Moon takes 27 days 7 hours 43 minutes 11.5 seconds (27.321661 days) to complete one orbit of Earth. This is known as the sidereal month. Unfortunately for students, however, this is just one of five kinds of astronomical month. The anomalistic month is when the Moon goes from perigee to the next perigee. Since the line connecting perigee and apogee (called the line of apsides) advances along the orbit, completing a complete circuit in about nine years, this month is a bit longer, 27 days 13 hours 18 minutes 33.2 seconds, or 27.554551 days.

The nodal (or draconic) month is based on the Moon returning to the same point in space where it crossed the intersection with Earth's orbit. Since this intersection regresses around the Moon's orbit, it takes only 27 days 5 hours 5 minutes 35.8 seconds, or 27.21220 days. A complete regression all the way around the Moon's orbit takes 18.6 years, and this relates to the frequency of eclipses, discussed below.

The tropical month is based on the Moon's position with reference to the vernal equinox, and takes 27 days 43 minutes 4.7 seconds, or 27.321582 days.

All the above are a bit short for a calendrical month, and in fact no culture ever based a calendar on any of the above, although some ancients, such as the Babylonians, and presumably the people who built Stonehenge, were aware of the nodal month, since they could predict eclipses, which are based on the nodal month. The calendar's months were originally based on the synodic month of 29 days 12 hours 44 minutes 2.9 seconds, which is the time period for phases to repeat. It is suggestive that the word synodic derives from an ancient Greek word for a meeting of priests. This month is 29.530589 days in average length, but because of minor gravitational effects from the Sun and a few planets, can be as short as 29.18 days and as long as 29.93 days. For this reason, occasionally February will lack one of the phases, although January and March will then have the missing phase twice.

The Moon's orbit was known to an accuracy of about a mile as recently as the 1960s. However, four of the Apollo landings placed a laser reflector on the Moon, as did an unmanned Soviet landing. This has permitted defining the Moon's position to an accuracy of one centimeter.

ECLIPSES

There are two main types of eclipse, lunar and solar. Both types require that the Moon be near one of the two intersections of its orbit with that of the Earth. These two points are called the nodes, one rising, where the Moon goes from below the plane of Earth's orbit to above, and the other the descending node where it is dropping below the plane of Earth's orbit. Since eclipses can only happen when the Moon is near this plane, it is known as the ecliptic. The Sun is always on the ecliptic, while the planets have orbits tilted to the ecliptic, and hence can usually be found near but not precisely on the ecliptic. Most months the Moon passes too far above or below the ecliptic or is in the wrong phase when at a node for an eclipse to occur.

A solar eclipse can only happen when the Moon is directly between Earth and Sun. Its phase at that time is New, so the Moon is not visible except for any sunlight reflected off Earth and onto the Moon. Lunar eclipses happen when the Moon moves into Earth's shadow. This is possible only at Full Moon. The Moon's shadow cannot be larger than about 160 miles across at Earth's distance, so the shadow creating a solar eclipse will never cover more than that at any instant. The shadow will move across the Earth's surface quickly enough that no solar eclipse can last longer than seven minutes fifty seconds, and most are just a minute or two in duration. NASA has posted on its website links to all types of eclipses from 2000 B.C. to 3000 A.D.

It is one of the oddities of the Solar System that our Moon appears almost exactly the same size as the Sun, and so it just blocks the Sun in solar eclipses, leaving the Sun's atmosphere, the corona, visible around the dark shape of the Moon. This is the only time it is safe to observe the Sun with unprotected eyes. At all other times, including during partial solar eclipses, enough Sun is visible and bright enough to cause eye damage. Galileo helped prove this by studying the Sun long enough to spend the last twenty years of his life nearly or totally blind. (It is virtually certain that when Galileo was subjected to the second degree of torture by the Inquisition, which involved being shown the instruments of torture while listening to the dungeon master gloat over their use, he saw nothing, dungeons generally being ill-lit.)

In a lunar eclipse the Moon moves into Earth's shadow, which blocks most sunlight from reaching the Moon. Our atmosphere allows some light to get around the edges of the Earth, and since this means effectively that it is the twilight glow that gets through, the eclipsed Moon usually is a dim object with the distinctly reddish tint of twilight. Earth has a diameter nearly four times that of the Moon (7927 miles versus 2159.9 miles), so the entire Moon can be inside Earth's shadow for up to two hours. When the Moon is completely inside Earth's shadow we have a total, or umbral eclipse. The umbra is the central part of Earth's shadow where no sunlight (except the twilight effect already mentioned) reaches the Moon. If the Moon is partly inside the umbra, we have a partial eclipse, and from Earth we see part of the Moon as a darkened, reddish object, and the rest looking almost normal. However, that outer part of the Moon is in a region called the penumbra, where an astronaut on the Moon would see the Sun partially covered by Earth. Penumbral eclipses are when the Moon is only in the penumbra, and not at all in the umbra. Penumbral eclipses are almost unnoticeable.

Solar eclipses occur slightly more often than lunar, but because at least half the Earth can see any lunar eclipse, while only limited areas see any solar eclipse, far more people see the lunar variety. Any calendar year must have at least two solar eclipses, and can have as many as five. There can be years with no lunar eclipse, or up to four. The maximum number of combined eclipses in any year is seven.

From 2000 B.C. to 3000 A.D. there are 12,186 lunar eclipses of all types, broken down into 4468 penumbral (36.7%), 4213 partial (34.6%), and 3505 total (28.8%).

If the Moon is near apogee at the time of a solar eclipse, and Earth is near its closest to the Sun, the Moon does not quite completely cover the Sun. In this case we can see a ring of the Sun around the Moon. This is called an annular eclipse, from the Latin word for ring, annulus.

Lunar eclipses are not known to have much effect on animal life, but solar eclipses have a big effect. As the sky darkens I have seen a herd of cows head for their barn, and in confusion reverse direction a few minutes later when the eclipse is ending. Daytime birds and insects fall silent in the last two minutes before totality, when the world darkens with an eerie gray tone. As totality falls, nighttime birds and insects awaken. The minute or so of silence adds to the weird effect, and undoubtedly contributed to the superstitious fear of eclipses. Most interesting was a dog taken to two eclipses. At the first it whimpered in fear, its tail between its legs, hiding under its owner's truck. At the second eclipse it was an absolute pest to everyone in the area, trying to play.

The motion of the nodes, where the Moon's orbital plane crosses the plane of Earth's orbit, controls not only the frequency of eclipses, but also causes a repetitive pattern called the saros cycle. This cycle was known to the ancient Greeks, from whom the name is taken, but they apparently learned of it from the Babylonians, and the cycle is built into Stonehenge as well. Any particular eclipse will be repeated 18.6 years later. The differences will be that its path is moved about 120 degrees westward across Earth's surface, and will move slightly north or south from the path of the previous eclipse. Thus I saw a 1962 solar eclipse in Maine, and the next member of its saros cycle just outside Irkutsk, on the shore of Lake Baikal.

SURFACE

The lunar surface was first studied in detail by Galileo Galilei (1564-1642), using a refractor telescope he had built himself based on a description mailed to him by a former student who had seen one of Hannes Lippershey's telescopes in Paris. Galileo noted indentations in the lunar surface, which he named craters (the Latin word for cup). The large, dark, flat areas he at first thought were lunar seas and oceans so he named them mare, plural maria, the Latin word for sea. The light colored regions he called highlands. All this terminology is still used, not only for our Moon, but for other moons, and even planets where appropriate.

In the mid Twentieth Century there was debate over the origin of the craters, however today we know nearly all are the result of meteors or asteroids striking the Moon. The largest are Mare Orientale, about six hundred miles across, but only partially visible from Earth, as it sits right on the dividing line between the side facing us and the farside, and the Aitken South Pole Basin, also poorly placed for viewing from Earth. However, it is believed most of the other maria are also the results of impacts. Molten basaltic materials from the lunar interior are believed to have risen to the surface, creating the mare where these large impacts occurred. Deep below the surface of some large craters and maria there are dense regions called mascons, believed to be remnants of the impacting bodies which have managed to maintain their identities.

The typical crater will have a slightly raised rim, materials ejected from the crater. Sloping walls will lead to a reasonably flat floor. Material falling back into the crater will have buried the impacting object, assuming it did not totally fragment upon impact. Craters more than sixteen miles in diameter can have a bump, known as a central peak, at or near the middle. This is true not just for the Moon, but for Earth and any other object large enough to have craters of more than sixteen miles. This provides a crude but effective way to estimate the sizes of craters on objects where central peaks can be found.

The maria were chosen for the early Apollo landings, since NASA was seeking the safest and easiest landing spots. It turned out that the local rocks contained a number of minerals not known on Earth, most prominently one abbreviated as KREEP, for potassium (chemical symbol K), Rare Earth Elements, and Phosphorus. The rocks are all of igneous origin, with some metamorphosed. They run around 3.6 to 3.9 billion years old.

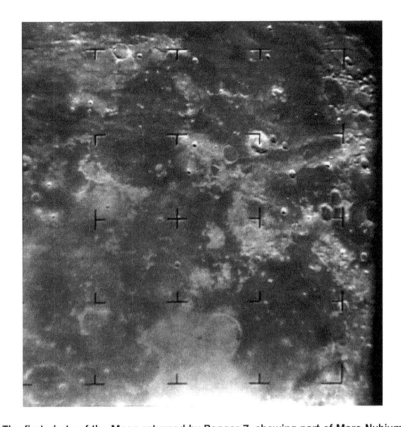

The first photo of the Moon returned by Ranger 7, showing part of Mare Nubium

The lunar highlands are of a lighter colored rock called anorthosite, rich in magnesium and aluminum. The highlands are older, about 4.3 billion years, and are believed to be the remnants of the Moon's original crust.

The Moon is also known to have what seem to be a few small extinct volcanoes, as well as lava tubes. These tubes in some cases run for dozens of miles. They were formed when lava flows cooled on their surface, but remained fluid underneath. The lava poured out, leaving a hollow tube. Some are a couple hundred yards wide, and would make excellent shelters from solar flares or meteors, should future bases on the Moon find this necessary.

The near side of the Moon is more than half covered with maria, while there is no true mare on the farside. This is believed to relate to the Moon's origin. The current popular theory among astronomers is that Earth was struck by an object roughly the size of Mars when the Solar System was quite young. Much of the impacting object merged into Earth, but a lot was ejected back into space, forming at least two bodies. Eventually the smaller one was pulled into the larger. The rough and mare-free farside is where the smaller one smashed into the Moon.

Our Moon has a density of 3.32, where water is 1, and Earth has an overall density of 5.52. This makes it one of the densest moons.

SPACECRAFT

The first spacecraft to try to visit the Moon was launched by the old Soviet Union. It was named Luna 1 by the Soviets, but it and its successors were often called Luniks in the West. The following table shows some of the major successful spacecraft to visit the Moon.

Spacecraft	Date Launched	Comments
Lunik 2 (USSR)	Sept. 12, 1959	First to impact on Moon, Sept. 14, 1959
Lunik 3 (USSR)	Oct. 4, 1959	First to photograph farside
Ranger 7 (USA)	July 28, 1964	close up photos, impact July 31, 1964
Lunik 9 (USSR)	Jan. 31, 1966	First soft landing (on Oceanus Procellarum)
Lunik 10 (USSR)	March 31, 1966	First satellite of Moon April 3, 1966
Surveyor 1 (USA)	May 30, 1966	soft landing (on Oceanus Procellarum)
Lunar Orbiter 1 (USA)	August 10, 1966	First extensive mapping
Apollo 11 (USA)	July 16, 1969	First manned landing (on Mare Tranquillitatis), July 20, 1969
Lunik 16 (USSR)	March 12, 1970	sample return mission
Lunakhod 1/Lunik 17 (USSR)	Nov. 10, 1970	automated rover
Hiten (Japan)	Jan. 24. 1990	mapping
Clementine (USA)	Jan. 25. 1994	search for polar ice
SMART-1 (European Space Agency)	Sept. 27, 2003	mapping, radiation
Kaguya (Japan)	Sept. 14, 2007	geology, origin, gravity field
Chang'e 1 (China)	Oct. 24, 2007	3-D mapping
Chandrayaan 1 (India)	Oct. 22, 2008	surface chemistry and polar ice
Lunar Reconnaisance Orbiter (USA)	June 18, 2009	topology, high resolution mapping

The USSR allegedly tried about twenty unmanned spacecraft to the Moon. About half failed. The two sample return missions brought back less than six ounces total of lunar soil.

The USA failed with five Pioneer spacecraft launched at the Moon, as well as six failures with Ranger spacecraft before the last three were able to take close up photographs of the lunar surface prior to impact. These pictures demonstrated that the Moon has craters down to the minimum size the spacecraft could record, about sixteen inches. There were seven Surveyor spacecraft, whose prime goal was to prove safe landings were possible. Surveyors 2 and 6 failed. Apollo 12 landed near Surveyor 5 and retrieved some parts of it for study of how the materials survived their time on the Moon. The Lunar Orbiters were intended to provide complete maps of the Moon in support of the planned Apollo landings, and the five were the most successful space program ever, as all five worked perfectly. NASA sent film strips of all the photos from the five to the libraries of all the colleges in the United States, and to many other places.

There were nine Apollo launches to the Moon, of which two (#8 and 10) were test flights not intended to land, six landed, and one (#13) experienced difficulties which prevented a landing. Apollo missions 18 through 20 were cancelled by the Nixon Administration, and the funds used for the VietNam war. A couple fiction novels and at least one movie have claimed additional Apollo missions, which never existed in reality, e.g. *The Cassandra Project*, by Jack McDevitt and Mike Resnick.

Apollo 12, which landed close to the Surveyor 5 unmanned probe,
and retrieved portions for study of how exposure to space conditions affected it

MARS

Mars has two moons, Phobos and Deimos. (For all planets with more than one moon, this book takes them up in the order going outward from the planet.) Both moons were discovered at the U.S. Naval Observatory in Washington, D.C. by a husband and wife team using a new Alvan Clarke telescope. Asaph Hall (1829-1907) was the new director of the observatory. Angelina Stickney Hall (1830-1892) had been his teacher, and did the necessary mathematics for the two of them. Both moons have synchronous rotation, meaning they rotate in the same amount of time that they revolve around Mars.

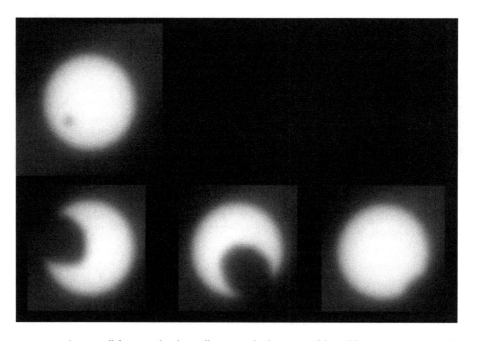

The martian moons are too small for a real solar eclipse, producing something akin to an annular eclipse or transit. Upper photo, solar transit by Deimos; lower photos a sequence showing a solar transit/eclipse by Phobos

The names for both moons were suggested by a teacher in Britain, and are taken from the Illiad, where Mars hitches up his horses Phobos (panic) and Deimos (terror) to his chariot and heads off for battle. The English word phobia is derived from the Greek.

[M 1] Phobos

Phobos was discovered on August 18, 1877. Initially the Halls believed it might be two or three moons, as no one had ever imagined a moon going around its planet in less than one planetary day. Phobos goes around Mars in 7 hours 39.2 minutes, while a Martian day is 24 hours 37 minutes. This means Phobos will go around Mars 3.2 times in one day, and therefore will rise or set as many as four times from some locations. If Phobos rises in the west as the Sun is setting there, it will go from New to waxing gibbous before setting in the east in less than four hours. If it is rising in the west as the Sun is rising in the east, its phase will be full, and in a few hours it will create a 20 second solar eclipse at New, and shortly after noon it will set in the east as a waxing crescent. Because Phobos is so close to Mars, it cannot be seen on the surface more than 70.4 degrees from the equator.

Phobos is in a low orbit, ranging from 5738.7 miles (the Martian equivalent to perigee is called periareion, or for universal simplicity periapsis) to 5914.7 miles (apapsis) above the center of Mars. (Mars has a diameter of 4260 miles, so deduct half of that to find how close it is to the surface.) Every time Phobos goes over the daytime side of the planet it causes a solar eclipse, although it is too small to completely cover the Sun, so it is more like the annular eclipses our Moon sometimes creates. Phobos also experiences a lunar eclipse every time it passes over the night side of Mars. For Earthlings, the apparent magnitude when Mars is at opposition is 11.4.

M1, Phobos from 4100 miles

Phobos has been found to have a low density of 1.876. The likeliest explanation it that it was disrupted into pieces one or more times, and as these coalesced back into a single body gaps were created. The surface area is only 240 square miles. This is less than a third the size of the District of Columbia, and close to the size of Brooklyn. The escape velocity is just 25 miles per hour, so while an astronaut could not leap off into an escape orbit, very long jumps are possible, and a launch ramp could be built for bicycles into space.

The material Phobos is made of is extremely dark, and probably has a lot of carbon. This makes Phobos resemble asteroids known as the carbonaceous class, and has led some to suggest it is a captured asteroid. There are major problems with this theory, as a capture would almost certainly result in a very elliptical orbit with a considerable tilt to the planet's equator, and Mars has such a weak gravity field that captures are difficult to achieve. Phobos has craters, and a system of grooves believed to be related to impacts cracking the surface. The surface has a layer of dust. Some craters are named for characters out of *Gulliver's Travels.* The book, written 151 years before the moons were discovered, says scientists of the flying island of Laputa had discovered Mars had two moons! Other craters are named for people who actually studied the moons, including the largest crater being named Stickney.

[M 2] Deimos

The Halls discovered Deimos on August 12, 1877. It is even smaller than Phobos, 9.3 by 7.6 by 6.4 miles. (Objects less than roughly 250 miles diameter are not pulled into a spherical shape by gravity.) This gives it an average radius of 3.9 miles. The density is even lower than that of Phobos, 1.471. The surface gravity is 0.0004G, where G is the gravity on Earth's surface. Thus an astronaut plus spacesuit weighing a total of 300 pounds on Earth would weigh 0.12 pound, or just about two ounces, on Deimos. This is a fine example of the difference between weight and mass. Deimos lacks a layer of dust on the surface, probably because any impact shakes things up enough to shake loose material off into space. The albedo is 6.8. The apparent magnitude is 12.45 at mean opposition.

Deimos orbits 14,580 miles from the center of Mars, in a period of 30 hours 18 minutes, or 1.26244 days. This is so close to the time Mars takes for a day that during the time between rising and setting for Deimos from any position on Mars 2.7 martian days will elapse. Thus if Deimos is rising in the east and Full at sunset, it will slowly crawl across the sky with the phase going through waning gibbous to just past Third Quarter at sunrise. A few hours later it will be New, and can transit the Sun.

Deimos has a visual diameter of only 2.5 minutes of arc from Mars, making it no more than a dot as bright as Venus looks in Earth's sky to most people, although a few with exceptional vision might barely be able to see a body. It will not be visible from any location within 7.3 degrees of the Martian poles. Mars will be 40 times larger from Deimos than the Full Moon looks to us.

Deimos has two craters named Swift and Voltaire. Swift is for Jonathan Swift (1667-1745), author of *Gulliver's Travels,* and Voltaire (1694-1778) because he had a visitor from Sirius in *Micromegas* mention the two moons of Mars, again, well before they were discovered. The reason for these predictions is probably attributable to Kepler, who noted that Earth had one moon, and Galileo had found four around Jupiter, so it seemed reasonable to expect the planet between them to have two.

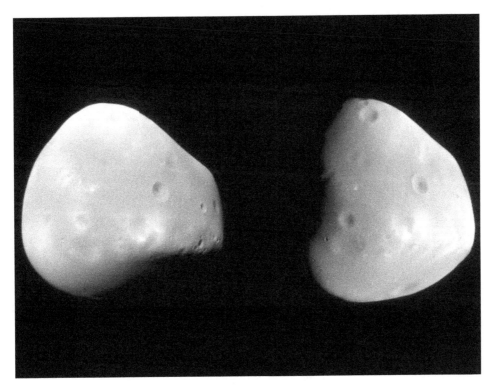

Deimos, M2, from 320 miles

SPACECRAFT

The following spacecraft have photographed Phobos or Deimos from space:

Mariner 9	1971	both
Viking 1 Orbiter	1977	both
Phobos 2	1989	Phobos
Mars Global Surveyor	1998	both
Mars Express	2004	both
Mars Reconnaissance Orbiter	2007	both

From the surface of Mars:

Opportunity	2004	both
Spirit	2005	Phobos

JUPITER

The latest count gives Jupiter at least 67 moons, which are regarded as falling into several groups plus a few moons that fall into no grouping. It is likely Jupiter has another twenty or more as yet undiscovered very small moons.

INNER GROUP

[J 1] Metis

In Greek mythology Metis was the first wife of Zeus, a goddess of wisdom, and mother of Athena. Thus it is appropriate that the moon nearest Jupiter (the Roman equivalent of Zeus) be given this name. It was discovered on photographs returned in 1979 by Voyager 1 as that robot flew past Jupiter. Eagle eyed Stephen Synnott spotted the faint dot on the transmissions. In 1996 the Galileo probe managed to get some better pictures. Metis is 37 miles by 24 miles by 20.5 miles, with the long axis pointed to Jupiter, and like many moons, the same face is always pointing to its planet. It is 77,988 miles from Jupiter's center, with an orbit so nearly circular that this varies by only 15 miles either way. The orbit is tilted by only 0.06 degree from Jupiter's equator. This moon is magnitude 17.5 at Jupiter's mean opposition.

Metis has such a low density that it is believed to be made almost entirely of ices. (To astronomers the frozen forms of water, carbon dioxide, ammonia, sulfur dioxide, and methane are ices. Jupiter is too close to the Sun for its moons to have frozen methane.) The lead hemisphere as Metis goes 19.56 miles per second around Jupiter is distinctly brighter than the trailing hemisphere. (Thus Metis moves around Jupiter one mile per second faster than Earth goes around the Sun.) This is assumed to result from impacts stirring up ices from the inside. The escape velocity is about 400 feet per second.

[J 2] Adrastea

This moon was discovered by David Jewitt (1958-) on photographs from Voyager 2 on July 9, 1979. It is 80,100 miles from the center of Jupiter on an orbit tilted only 0.03 degree to Jupiter's equator. It takes 7 hours 9.5 minutes (0.29826 day) to orbit the planet, and was therefore the third moon known to orbit a planet in less time than the planet takes to rotate. It is 12.5 by 10 by 8.5 miles, with the long axis pointed to Jupiter in sychronous rotation. Surface gravity would be 0.00004G, with an escape velocity of 150 feet per second. It orbits near the outer ring of Jupiter, and is assumed to be the source of material for the ring as meteors pound on Adrastea's surface.

Adrastea is magnitude 18.7 when Jupiter is at mean opposition.

Adrastea was a mountain goddess originally from Phrygia noted for defending the righteous.

[J 3] Amalthea

Amalthea was the infant Jupiter's nursemaid. When he was nine months old he seduced her, which suggests Greco-Roman gods were rather precocious. This moon was discovered by Edward Emerson Barnard (1857-1923) on September 9, 1892 working at Lick Observatory, the first Jovian moon to be discovered since Galileo, and the last moon anywhere to be discovered visually rather than with the modern use of photography. Barnard disapproved of using classical mythology for the names of moons, and insisted on calling it simply Jupiter V. The French astronomy populizer Camille Flammarion proposed the name Amalthea, although it was not officially adopted by the International Astronomical Union until 1975.

Amalthea is 180 miles by 90 miles by 80 miles, one of the largest nonspherical objects. It orbits 112,400 miles from Jupiter's center, with a synchronous period of 11 hours 57 minutes 23 seconds (0.49817943 day). It spends 90 minutes in Jupiter's shadow each orbit. From Amalthea Jupiter would spread across 46 degrees of the sky, 92 times the size of the Moon in our sky. Amalthea spends 29 hours going across Jupiter's sky because of the similarity of its orbital period to the length of a Jovian day.

Amalthea has an albedo of 9, and is apparent magnitude 14.1 at opposition. The surface is very red except around two large craters. It is believed the red surface comes from sulfur ejected by Io's volcanoes, while the craters are fairly recent and their floors not yet coated. The craters are respectively fifty and sixty miles in diameter, five to ten miles deep. Considering the size of Amalthea, and looking at large craters on our Moon and Callisto, it seems that quite large objects can collide without being disruptive.

[J 4] Thebe

Thebe is named for a Greek nymph who had an affair with Zeus and later married Zethus. The only named feature on the moon is a crater named Zethus. Thebe was discovered by the ever vigilant Stephen Synnott on photographs taken by Voyager 1 on March 5, 1979. It was the fourteenth moon of Jupiter to be discovered.

Thebe takes 0.674536 day, or 16 hours 4 minutes 17 seconds to orbit Jupiter. It is ranges from 132,000 to 136,900 miles from Jupiter in an orbit inclined 1.08 degrees to the planet's equator. It is red and has an albedo of 4.7, almost certainly deposits from Io's volcanoes releasing sulfur into space. Its dimensions are 71.5 miles by 61 miles by 52 miles, with the crater Zethus 25 miles across. Thebe is magnitude 16.0.

GALILEAN

Galileo Galilei (1564-1642) turned his telescope on Jupiter in early January 1610, and on January 7 discovered the first three moons, Io, Europa and Callisto. On January 10 he discovered Ganymede. Galileo indicated a desire to name the moons collectively Stellae Mediciorum, or the Medici Stars. The Medicis were not universally admired, and the choice fell flat. Simon Marius independently discovered the moons a few days after Galileo (or so he claimed), and his suggested names are the ones that were adopted. Io was a nymph and priestess seduced by Zeus. Europa was carried off to the island of Crete by Jupiter in the guise of a bull. Ganymede was a pretty boy Jupiter took to Mount Olympus for lustful purposes. After he tired of Ganymede, the boy served as cupbearer, the only part of the legend mentioned by those opposed to gay rights. Callisto can be translated either as cutey or blondie.

[J 5] Io

Galileo's first view of Io in his poor quality homemade telescope on January 7, 1610 could not separate the images of Io and Europa. That had to wait until the next day. He reported the discovery, along with a number of others, in his book *Sidereus Nuncius,* published later that year.

Io has a diameter of 2263.4 miles, making it the fourth largest moon in the solar system. It also has the highest density of any moon, 3.5275. This suggests a core of iron or more likely iron sulfide. Io takes 1.769137786 days (42.45930686 hours, or 1 day 18 hours 27.6 minutes) to orbit Jupiter at a distance of 260,800 to 262,900 miles. Io's orbit lies within Jupiter's Van Allen Belt, a region where the planet's magnetic field captures protons, electrons and other charged particles from the Sun. Particles in the Van Allen Belt can strike Io's surface. Jupiter's magnetic field is skewed dramatically with respect to its orbit or equator, so Io can pass above or below the Van Allen Belt at times during its orbit. However, when it is within the Belt the surface will be exposed to radiation of as much as 3600 rem per day. A count of over 1000 rem in a day is invariably fatal to humans and most other terrestrial life forms. The magnetic field also contributes to giving Io auroras.

The orbit is inclined 0.05 degree to Jupiter's equator. Were Io not so close to Jupiter as to be generally obscured by the planet's much greater brightness, it would be visible at opposition to the unaided eye, with a magnitude of 5.02. Io is noted for having more than 400 active volcanoes, much more than any other object in the Solar System, including Earth. The volcanoes were discovered on photographs sent back by Voyager 1 by Linda Morabito (1953-). The volcanoes are believed to be dependent on a magma ocean thirty miles below Io's surface, and extending down another 30 miles. It would take up ten percent of the mantle and have a temperature of 1200C.

Io also has close to 150 mountains which average four miles high, with the highest going up to eleven miles. (Contrast Earth, where Mount Everest is 5.5 miles high.) Mountains are most common in regions with few volcanoes. The reverse is also true. It is believed that the mountains result from compressional strains on Io's crust.

Io has a very thin atmosphere composed of sulfur, oxygen, chlorine, sodium, potassium, and sulfur dioxide. This last is not far from its freezing point, and produces an interesting effect. Io is slightly brighter when it comes out of an eclipse in Jupiter's shadow than at other times, and the brightness fades to a normal appearance within twenty minutes of leaving eclipse. It is believed that the somewhat chillier temperature in Jupiter's shadow causes the sulfur dioxide to freeze out and coat the surface with a thin layer of SO2 ice, which evaporates back to a gas when the eclipse ends. SO2 ice is found on the surface of Io near the poles at all times.

Io's volcanoes are caused by the gravity of Jupiter, Europa and Ganymede pulling on Io, shifting its shape. The friction generated creates heat resulting in melting some volatile materials. The main material affected is sulfur. Sulfur has the unusual characteristic of being different colors depending on how rapidly the molten form cools to a solid. With lava flows of mostly sulfur covering over half of Io's surface area, it is noteworthy that the surface colors include yellow, orange, red, and black. When the first color photos of Io were received, someone commented that it looked like a diseased pizza. The volcanoes are concentrated within thirty degrees of the equator. Hot gases are ejected to altitudes of a couple hundred miles above the surface, and some of the sulfur vapor winds up escaping into space, to be dragged towards Jupiter by its gravity. This is the source of the colorful sulfur coating on the inner moons. On Earth lava runs around 2200F. Io is nowhere near that hot, so its lavas are not molten silicates, but primarily molten sulfur, which melts around 660F. Gravity pumping is also found on several other moons, although melting water ice rather than sulfur.

The cover of this book has a color photograph of Io, showing volcanoes and other surface features.

[J 6] Europa

Europa is 412,400 to 420,380 miles from Jupiter, in a slightly elliptical orbit. It takes 3.551181 days, or 3 days 13 hours 13.7 minutes, to orbit Jupiter. The orbit is tilted 0.47 degree to Jupiter's equator. The diameter is 1961 miles, with a surface area 6% of Earth's. Since Earth has a surface that is 70% water, the land area of Earth is five times the surface of Europa. It has a very high albedo of 65, and with a visual magnitude of 5.29, it would be visible to the unaided eye if Jupiter's glare were not hiding it.

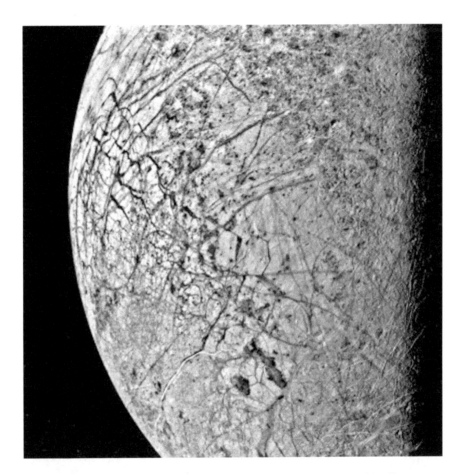

Europa, J6, showing lines of fractures and ridges from 417,489 miles

Europa lies near the outer part of Jupiter's Van Allen Belt, with a reading of 540 rems per day on the surface. The surface is the smoothest of any solid object in the solar system. Few craters can be identified, and much of the surface shows little but shallow ridges or grooves. The surface has dark reddish streaks which seem to be sulfur and magnesium sulfide. There is an extremely thin atmosphere (about 10^{-12} the density of Earth's), mostly of oxygen. While on Earth the oxygen is a sign of the presence of life, on Europa the oxygen results from solar radiation breaking up water molecules. Europa may have a metallic core, but most of its matter is water ice, with the surface an ice crust whose thickness is disputed by astronomers and geologists. Some believe it is at most a couple miles thick, while others say sixty miles is more likely. Several experiments for future spacecraft have been proposed to resolve the question, as, if the crust is thin, under it would be an ocean of liquid water (possibly salt water), perhaps capable of sustaining some form of life.

Flexing from the gravity of Io and Ganymede produces internal friction. This friction in turn creates heat, which would raise internal temperatures enough to create the subsurface water. This need not be one large continuous body, but could be broken up into separate seas, each of which could have its own life forms.

[J 7] Ganymede

Ganymede is the largest moon in the Solar System, and in fact is larger than the planet Mercury, 3273 miles versus 3010 miles. It is 663,980 to 665,420 miles from Jupiter, taking 7 days 3 hours 42.6 minutes, or 7.15455296 days, to orbit the planet. The orbit is tilted just 0.20 degree to Jupiter's equator. The density is 1.936, suggesting a fairly even mix of ices and rocky material. Since it also has its own magnetic field, the only moon to do so, it must have an iron core of some size, and can have auroras down to 50 degrees from either pole. The core could be or include iron sulfide. The core would be about 1000 miles across, and overlaid with a 500 mile mantle of olivine and pyroxene. The remainer would be primarily ices. This means that Ganymede has separated its materials, with the most dense settling to the center and least dense for the surface crust. This process is known as differentiation. Earth is highly differentiated also. A high degree of differentiation generally suggests an object was totally molten at some point in its history.

With an albedo of 43, Ganymede has a visual magnitude of 4.6 at opposition. This would be easily visible without the planet's glare, and in fact there are cases known of people seeing Ganymede without any visual aids, usually using a tree limb or building to block Jupiter. China claims a third century B.C. Chinese astronomer may have recorded Ganymede.

Ganymede, J7, the solar system's largest moon, showing detail down to 9.5 miles across

The various spacecraft that have visited the Jovian system have found that Ganymede's surface is broken up into large plates, similar to Earth's tectonic plates. These plates have even shifted the way Earth's do, producing fault lines, buckling, hills, and other surface features. The surface is more than 50% water ice, with CO_2, sulfur dioxide, cyanogen, hydrogen sulfide, magnesium sulfate and sodium sulfate. The oldest parts of the surface are darker than younger regions, and are saturated or close to saturated with craters. Saturation means that for every new crater produced by an impact, one old crater gets obliterated. The younger regions are the areas where grooves and ridges are found, but these both cross craters and have craters imposed on them.

Unlike the moons closer to Jupiter, all of which have leading hemispheres that are brighter than the trailing hemisphere as they orbit the planet, Ganymede's trailing hemisphere is brighter. It is also enriched with sulfur dioxide.

Ganymede is the main human colony in the Jovian system according to science fiction author Robert A. Heinlein (1907-1988), and most of his novel *Farmer in the Sky* is set there. It is mentioned in several of his stories, and other writers have also based stories on Ganymede. This of course was written before the high level of radiation from Jupiter's Van Allen Belts was discovered.

[J 8] Callisto

Callisto is 1,161,397 to 1,178,796 miles from Jupiter. This takes 16.6390184 days, or 16 days 16 hours 32.2 minutes. Callisto's orbit is inclined with respect to Jupiter's equator by 0.192 degree. It has a density of 1.83. Evidence suggests that Callisto is partially but not completely differentiated. The diameter is 2995 miles, with an escape velocity of 1.52 miles per second. Surface gravity is 0.126 that of Earth.

Callisto's albedo is 22, and it reaches magnitude 5.65 when Jupiter is at opposition. This is barely bright enough to see, and Callisto when at its maximum east or west of Jupiter (maximum elongation in astronomical jargon) should be visible with difficulty. The surface gets less than 0.2 rem per day of radiation from Jupiter's Van Allen Belt, making it the only Galilean moon suitable for long term human landings.

In contrast to most other moons near Jupiter, Callisto's trailing hemisphere is brighter than the leading one, with dry ice (frozen carbon dioxide) predominating. The lead hemisphere has sulfur dioxide ice dominant. The surface is very old and is saturated with craters. Several mounds look as though they may be inactive or rarely active cryovolcanoes. A cryovolcano erupts as molten things that on Earth would not be volcanic, such as water.

Callisto has two enormous craters, Valhalla Basin at 1100 miles diameter and Asgard Basin at 990 miles. As with nearly all Callistan craters, the rims are very low and the craters very shallow, suggesting that the mantle or layer beneath is mostly ice and unable to bear much weight. Spacecraft have identified water ice on crater rims, scarps and ridges.

Callisto has an extremely thin atmosphere of carbon dioxide.

The remaining moons are all small and very far from Jupiter. One of the consequences of their distance is that the Sun, and to a minor extent other planets, greatly influence their orbits, and the numbers given for distance, orbital period, and even to some extent orbital inclination change rather drastically. The numbers given below are typical, but not necessarily the precise values in effect when you read this.

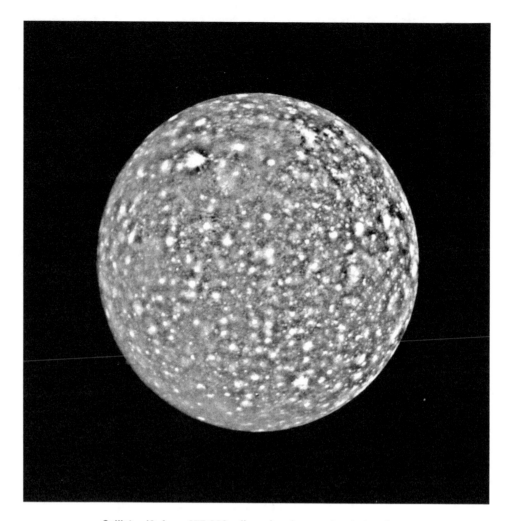

Callisto, J8, from 675,000 miles, showing a saturated surface.

[J 9] Themisto

Themisto is a lonely moon, not being a member of any of the groups other moons are part of. It is named for one of those nymphs Zeus/Jupiter was always pursuing, the daughter of a minor river god named Inachus. She was involved in a domestic tragedy due to jealousy over her husband's second wife.

Themisto was discovered by Charles Kowal (1940-2011) and Elizabeth Roemer (1929-) on September 30, 1975, and promptly lost until the year 2000 when two different groups of astronomers independently found it. It was the eighteenth moon of Jupiter to be discovered, and so is sometimes written as Jupiter XVIII.

Themisto ranges from 3,669,500 to 5,510,940 miles from Jupiter, with a fairly eccentric orbit. The orbit is inclined 47.48 degrees to Jupiter's equator. Compare that with the moons listed above, none of which even approach a one degree tilt. Themisto takes 130 days to orbit Jupiter.

Themisto has not been examined by any spacecraft, so the best estimate of its size is a five mile diameter. This would give it an albedo of just 4, since it is magnitude 14.4 at opposition.

HIMALIA GROUP

[J 10] Leda

Leda was discovered by Charles Kowal, working at Mount Palomar Observatory, on September 14, 1974. He initially called it Jupiter XIII. There is an absolutely atrocious 1950s movie about a spaceship visiting Jupiter's thirteenth moon, which was inhabited only by scantily clad "fire maidens" and some sort of monster whose sole goal in life seemed to be trying to kill them and the visiting Earth men. This was long before Kowal actually discovered a thirteenth moon. The name Leda was used in some of Robert Heinlein's science fiction for the main human colony on Ganymede.

Leda ranges from 5,785,600 to 8,110,000 miles from Jupiter, taking 240.92 days to orbit the planet. Its orbit is inclined 29.01 degrees to Jupiter's equator. It is only about six miles in diameter, with a density of 2.6. Surface gravity is one tenth of one percent of surface gravity on Earth, so those fire maidens could bounce around pretty good. It is magnitude 20.2 at opposition.

In mythology Jupiter, disguised as a swan, visited Leda. She later laid two eggs (!) which hatched and grew up to become Castor and Pollux (the Gemini twins).

[J 11] Himalia

Himalia is the largest member of this group, with a diameter of 83 miles. It was discovered by Charles D. Perrine (1867-1951) working at Lick Observatory on December 3, 1904. It is believed that all members of this group, which are characterized by similar orbits and a similar gray color, were at some time in the past a single object which was broken apart in a collision. A fifth member of the group, about one mile across, was photographed from Earth in 2000, and not recovered until 2012. Some dust seems to have been spread briefly along Himalia's orbit according to a photo from the New Horizons spacecraft.

Himalia is 7,116,600 miles from Jupiter on average, but swinging 1,007,000 miles closer or further than average in an elliptical orbit that is tilted 29.59 degrees to Jupiter's equator. It rotates on its axis in 7.78 hours, unlike the eight closest moons, which all have one face permanently towards Jupiter.

The Cassini spacecraft that flew past Jupiter in 2000 on its way to Saturn found Himalia to be gray, but with some indication of water ice being present. It was 2,700,000 miles away at the time, so this observation is not definite. New Horizons in 2007 was more than 5 million miles away when it took a photo of Himalia.

Himalia was a nymph from the Isle of Rhodes. She had three sons with Jupiter.

[J 12] Lysithia

Lysithia was discovered by Seth Nicholson (1891-1963) on July 6, 1938 at Mount Wilson Observatory. It is one of four moons of Jupiter discovered by Nicholson, putting him in a tie with Galileo. Lysithea was another one of those nymphs Jupiter/Zeus just couldn't resist.

Lysithia's elliptical orbit goes from 6,327,100 to 8,254,900 miles from Jupiter, in an orbit that takes 259.2 days, or 0.70965 year. Its orbit is tilted 25.77 degrees to Jupiter's equator. Lysithia has a diameter of eleven miles, and a density of 2.6.

[J 13] Elara

Elara was discovered by Charles Perrine on January 2, 1905. It is 5,889,360 to 8,738,960 miles from Jupiter with its orbit inclined 30.66 degrees to Jupiter's equator. Elara has a diameter of 53 miles. In February 2007 the New Horizons spacecraft on its way to Pluto photographed Elara and found that it rotates on its axis in 12 hours. Its orbital period is 259.64 days. The surface gravity is 0.003G, with an escape velocity of 170 feet per second. The apparent magnitude is +16.3.

Elara was one of Jupiter's many loves, and the mother of a giant in Greek mythology.

The International Astronomical Union has decreed that moons in this group must have names ending in the letter a. Jupiter seems to have had enough lovers to assure almost any letter can fulfill the needs.

[J 13.5] S/2000 J-11

This moon was discovered December 5, 2000 by Scott Shepherd, Marina Brozovic, and Robert Jacobson, but was lost after being tracked for just 26 days. Shepherd recovered it in July 2012, after a lengthy period when it was suspected that this moon may have impacted Himalia (J 11). It has a diameter of about 2.5 miles, a period around Jupiter of 287.93 days, and an inclination of 27.584 degrees. Its orbit ranges from 6,202,100 to 9,416,500 miles from Jupiter. The apparent magnitude is +23, explaining why it is hard to follow.

NO GROUP

[J 14] Carpo

Carpo has an unusual orbit which places it in a group by itself. It was discovered by Scott Sheppard in April 2003 at Mauna Kea. Carpo is 7,735,000 to 13,558,950 miles from Jupiter in an orbit inclined 55 degrees to Jupiter's equator. It takes 458.625 days to orbit the planet. The diameter is barely two miles. It is also Jupiter XLVI.

In mythology Carpo was a daughter of Jupiter and a nature goddess ruling autumn.

[J 15] S/2003 J-12

This, perhaps tied for smallest moon of Jupiter, with a diameter of less than a mile, was discovered by Scott Sheppard March 7, 2003. It also is in a group by itself. The orbit ranges from 7,324,000 miles to 14,886,000 miles from the planet, taking 489.72 days for an orbit. The inclination to Jupiter's equator is 143 degrees. An inclination of 90 degrees would go directly over the poles. Inclinations greater than 90 travel around the planet clockwise (called retrograde), which is the opposite of the majority of traffic throughout the Solar System. From this satellite on outward, all Jovian moons have retrograde orbits.

This moon has an absolute magnitude of 17.2. No name has been assigned.

ANANKE GROUP

All retrograde moons get names ending in >e<.

[J 16] Euporie

Euporie was discovered by Scott Sheppard in 2001, getting the designation Jupiter XXXIV. It is 10,715,420 to 12,991,380 miles from Jupiter, taking 538.780 days to go around once. The orbit is tilted 145 degrees to Jupiter's equator. It is only 1.3 miles in diameter.

Euporie (you poh ree) was the goddess of abundance and a daughter of Jupiter.

[J 17] S/2003 J-3

This Sheppard discovery has not yet received a name. The orbit's low point is 9,131,100 miles from Jupiter, and swings out to 15,239,100 miles, with an inclination of 146 degrees. It takes 561.518 days to orbit the planet. The diameter is about 1.3 miles.

[J 18] S/2003 J-18

Brett Gladman (1966-) discovered this as yet nameless moon in April 2003. The semimajor axis of this eccentric orbit is 12,303,600 miles from Jupiter in an orbit tilted 149 degrees. The orbital period is 569.518 days.

NO GROUP

[J 19] S/2011 J-1

This is one of the most recently discovered moons, found by Scott Sheppard in 2011, and reported in February 2012. It is only 0.6 mile in diameter, and takes 580 days to orbit the planet, with an inclination of 162.8 degrees. Its elliptical orbit takes it from 9,130,314 to 15,239,936 miles from the planet.

ANANKE GROUP

[J 20] S/2010 J-2

Christian Veillet discovered this so far unnamed moon in 2010. He was using the CFH 3.6 meter telescope in Hawaii. It ranges from 8,739,200 to 16,486,200 miles from Jupiter in an orbit inclined 150.4 degrees. It takes 597.607 days to go around Jupiter, and is only 1.25 miles in diameter.

[J 21] Thelxinoe

Thelxinoe was discovered by Scott Sheppard in 2003, and recorded as J XLII. It is only 1.25 miles in diameter, and goes from 9,392,600 to 16,111,000 miles from Jupiter with an inclination of 153 degrees. Thelxinoe takes 597.607 days to orbit the planet. It is named for a Muse who was also Zeus's daughter.

[J 22] Euanthe

Euanthe was the mother of the three Graces. Scott Sheppard found this moon in 2001, labelling it Jupiter XXXIII. It orbits 10,451,700 to 15,677,500 miles from Jupiter, taking 598.093 days for a complete orbit that is inclined 142 degrees to the planet's equator. Euanthe is only about two miles in diameter. The pronunciation is you an thee.

[J 23] Heliki

Heliki was discovered by Sheppard in 2003. It is called Jupiter XLV (bet you never expected astronomers to still be using Roman numerals). The orbit runs from 11,003,000 to 13,508,140 miles, taking 601.402 days to go around the planet. The orbit is inclined 156 degrees. Heliki is about 2.5 miles in diameter. It is named for a nymph who helped care for the infant Zeus. The name is pronounced heh lee kee.

[J 24] Orthosie

Scott Sheppard found this one in 2001, calling it Jupiter XXXV. It is 1.25 miles in diameter and 9,665,200 to 15,880,400 miles from Jupiter. It takes 602.619 days for an orbit that is tilted 143 degrees. Orthosie was the goddess of prosperity.

[J 25] Iocaste

Iocaste is one of several Jovian moons to share some form of a name with an asteroid, in this case Jokasta. Others include Ganymede, Ananke, Io, and Europa. Iocaste, whichever way you spell it, was the mother of Oedipus.

Scott Sheppard found this moon on November 23, 2000, and called it Jupiter XXIV.

Iocaste's orbit ranges from 10,358,500 to 16,051,400 miles from Jupiter, taking 609.427 days to go around. The orbit's inclination is 146 degrees to Jupiter's equator. This moon is about three miles in diameter, with a density of 2.6. The escape velocity is seven miles per hour. Compare the Earth, where the escape velocity is seven miles per *second*.

[J 26] S 2003 J16

Brett Gladman found this as yet nameless moon in 2003. It is 1.25 miles in diameter. The orbit ranges from 12,852,100 miles to 16,658,440 miles from Jupiter, taking 610.362 days to go around. The orbit is inclined by 149 degrees.

[J 27] Praxidiki

Scott Sheppard was responsible for finding this moon in November 2000. He called it Jupiter XXVII. Praxidiki was the goddess of punishment, the first stage of which no doubt was learning how to pronounce her name. Praxidiki takes 613.904 days to orbit Jupiter at a distance of 10,553,200 to 15,312,400 miles. The inclination is 143 degrees to Jupiter's equator (or 57 degrees moving backwards, if you prefer). It is believed to be about 4.5 miles in diameter, and to have a grayish tint.

[J 28] Hermippe

Hermippe was found by Sheppard in 2001 and designated Jupiter XXX. It is about 2.5 miles in diameter. It is named for a nymph who married Orchomenus, one of Jupiter's sons.

Hermippe averages 13,154,000 miles from Jupiter in an elliptical orbit that takes 629.807 days to complete. The orbit is inclined 149 degrees to Jupiter's equator.

[J 29] Mneme

Sheppard found Jupiter XL in 2003. It is only about 1.2 miles in diameter, and its elliptical orbit runs 8,963,400 to 17,279,800 miles from Jupiter. The orbit takes 640.769 days, and is inclined 148 degrees.

Mneme was a Muse; her father was Jupiter.

[J 30] Ananke

Seth Nicholson found Ananke while working at Mount Wilson Observatory in 1951. It was then recorded as Jupiter XII, Nicholson being of the old school which did not want to use names. The name is from the goddess of destiny, depicted holding a spindle, suggesting a relationship with the fates, especially Atropos.

Ananke is the largest moon in its group, with a diameter of 8.7 miles. Surface gravity is 0.1% the surface gravity on Earth. Ananke has a gray to light reddish color with a possibility of water ice.

The elliptical orbit swings from 7,804,100 to 18,048,500 miles from Jupiter, in 650.45 days (1.68 years). The inclination is 149.9 degrees.

[J 31] Harpalyke

Another Scott Sheppard discovery, from 2000, initially listed as Jupiter XXII. Harpalyke was another of Jupiter's love affairs. At 9,936,500 to 16,649,500 miles, it takes 654.542 days for an orbit, with an inclination of 147 degrees. The diameter is about 2.5 miles.

[J 32] Thyone

Thyone was another of Jupiter/Zeus's seemingly limitless lovers. Sheppard found this one in 2001, numbering it Jupiter XXIX. At an average distance of 13,293,000 miles, it goes around Jupiter every 659.803 days. The orbit is tilted 147 degrees to the planet's equator.

CARME GROUP

[J 33] Herse

Brett Gladman found this moon on February 8, 2003. It was listed as Jupiter L. Herse was a daughter of Jupiter and Selene, a Moon goddess. Her name means dew. Herse is believed to be only about 1.2 miles in diameter.

The orbit runs from 8,259,500 to 19,236,100 miles from Jupiter at an inclination of 165 degrees. It takes 672.752 days to orbit the planet.

[J 34] Aitne

Aitne was discovered by Scott Sheppard in 2001, and listed as Jupiter XXXI. It is just 1.3 miles in diameter, and runs 8,409,474 to 19,273,694 miles from Jupiter. The orbit is tilted 164 degrees. Aitne is a personification of Mount Etna, a well known Italian volcano. This moon is much too small to have a volcano.

[J 35] Kale

Another Sheppard discovery in 2001, listed as Jupiter XXXVII. It runs 11,117,500 to 16,114,500 miles from Jupiter, taking 685.324 days in an orbit inclined 166 degrees. It is only 1.2 miles in diameter. Kale was one of the Graces and a daughter of Jupiter.

[J 36] Taygete

Taygete shares a name with a star, one of the Pleiades also known as 19 Tauri. Taygete was a nymph who had a son, Lacodaemon, with Zeus. The name is pronounced tay juh teh. Scott Sheppard discovered this moon in 2000, recording it as Jupiter XX. Taygete is three miles in diameter, and orbits 8,809,300 to 19,059,600 miles from Jupiter. It takes 636.675 days for an orbit, which has an inclination of 163 degrees.

[J 37] 2003 J 19

This as yet nameless moon was discovered by Brett Gladman in 2003. It averages 14,102,300 miles from Jupiter, taking 699.125 days for an orbit, which is tilted 164 degrees.

[J 38] Chaldene

Another Sheppard discovery from 2001, Chaldene is Jupiter XXI. It has a diameter of 2.4 miles. The orbit ranges from 10,006,000 to 18,218,000 miles, taking 699.327 days. The tilt is 169 degrees.

Chaldene (kal dee nee) was one of Zeus's lovers.

ANANKE GROUP (AGAIN)

[J 39] S 2003 J 15

A still unnamed moon found by Sheppard in 2003. It has a diameter of 1.2 miles. The orbit is about 14,110,000 miles with an inclination of 142 degrees. The orbital period is 699.676 days.

CARME GROUP

[J 40] S 2003 J 10

Another unnamed moon from Sheppard in 2003. It is 1.2 miles in diameter, and averages 14,116,000 miles from Jupiter in an orbit tilted 166 degrees. It takes 700.129 days for an orbit.

ANANKE GROUP (RETURNS)

[J 41] S 2003 J 23

This nameless moon was on photos taken in 2003, but not noticed by Scott Sheppard until 2004. It is 1.2 miles in diameter, averaging 14,122,000 miles from Jupiter with a tilt of 149 degrees. It takes 700.539 days to orbit.

CARME GROUP (is back)

[J 42] Erinome

Scott Sheppard found this moon, Jupiter XXV, in 2000. It has a diameter of 2 miles. The orbit runs 10,638,474 to 17,928,856 miles from Jupiter with an inclination of 164 degrees. The orbital period is 711.965 days. It is named for another of Jupiter's many lovers. It is pronounced air ee no mee.

PASIPHAE GROUP

[J 43] Aoede

Otherwise Jupiter XLI, Scott Sheppard found this moon in 2003. Its diameter is 2.5 miles. The orbit runs from 8,141,000 to 22,451,000 miles from Jupiter with an inclination of 162 degrees. It takes 714.657 days to orbit. Aoede was a Muse and a daughter of Jupiter.

CARME GROUP

[J 44] Kallichore

This name is pronounced kal i kohr eh. Sheppard discovered this moon in 2003. It is also Jupiter XLIV. It is only 1.2 miles in diameter. The orbit is 11,423,100 to 17,281,775 miles from Jupiter with an inclination of 164 degrees. The orbital period is 717.806 days.

Kallichore was a nymph who cared for Dionysius.

[J 45] Kalyke

The pronunciation is kay lick ee. Scott Sheppard found Jupiter XXIII in 2000. The semimajor axis is 14,395,300 miles with a periapsis of 11,323,400 miles and an apapsis of 17,485,660. Kalyke takes 721.02 days to go around Jupiter with an inclination of 165.5 degrees. Its diameter is 3.2 miles. While all moons of the Carme group have a reddish tint, Kalyke is much redder than the others.

Kalyke was another nymph who cared for Dionysius, and may also have been one of Jupiter's lovers.

[J 46] Carme

Carme was discovered by Seth Nicholson using the 100 inch telescope at Mount Wilson in 1938. It became Jupiter XI. As the namesake of its entire class of moons and an early discovery, it is largest of its group, with a diameter of 28.6 miles, and a density of 2.6. Carme's orbit ranges from 11,032,078 to 17,779,800 miles. It takes 721.82 days to orbit Jupiter. The inclination is 167.53 degrees.

Carme was one of Jupiter's lovers and the mother of a Cretan goddess named Britomartis (not to be confused with a character in Spencer's *Faerie Queen*).

PASIPHAE GROUP

[J 47] Callirrhoe

Callirrhoe was photographed from Kitt Peak Observatory in October 1999, and initially reported as an asteroid. Timothy Spahr early in 2000 recognized it as a Jovian moon, now Jupiter XVII. It is 5.3 miles in diameter. The orbit is tilted 139.8 degrees to Jupiter's equator, and it orbits 10,701,000 to 18,150,500 miles from the planet. The orbital period is 722.62 days.

Callirrhoe was the daughter of a river god and another lover of Jupiter.

[J 48] Eurydome

Sheppard discovered this moon, Jupiter XXXII, in 2002. It is named for the mother of the Graces. Their father was Zeus. At 8,894,857 to 19,876,453 miles it takes 723.36 days to orbit Jupiter. The orbital tilt is 149.3 degrees. Eurydome is only 1.8 miles in diameter.

CARME GROUP

[J 49] S 2010 J1

Another so far nameless moon, discovered September 7, 2010 by Robert Jacobson, Marina Brozovic, Brett Gladman, and Mike Alexandersen. They were using the 200 inch Hale telescope at Mount Palomar. The smallest moon yet found, possibly less than 0.9 mile diameter, it takes 724.34 days for its 8,891,000 to 20,103,100 mile orbit. The inclination is 163.2 degrees. Alexandersen went over old photographs, and was able to find a very faint image on pictures taken in late February 2003.

[J 50] Pasithee

In 2002 Scott Sheppard found Pasithee, otherwise Jupiter XXXVIII. The 726.93 day orbit is 9,721,100 to 19,245,232 miles from Jupiter with a tilt of 165.8 degrees. It is 1.2 miles in diameter.

Pasithee was a member of the Charites (Graces) in charge of hallucinations and the substances that created them. Her father was Zeus.

NO GROUP?

[J 51] S/2011 J2

Another Scott Sheppard discovery, this one from 2011. It takes about 727 days to orbit Jupiter, and has a diameter of 0.6 miles. The inclination is 151.8 degrees. The elliptical orbit has a perijove of 9,845,100 miles and an apojove of 19,111,300 miles.

PASIPHAE GROUP

[J 52] Kore

Sheppard found Kore (pronounced kohr ee) in 2003. It is listed as Jupiter XLIX. It is 11,676,400 to 17,336,900 miles from Jupiter in an orbit that takes 727.72 days to complete. The inclination is 137.4 degrees. Kore is 1.2 miles in diameter. The name means daughter in ancient Greek, and stands for Persephone, daughter of Demeter.

[J 53] Cyllene

Sheppard also found this moon in 2003, actually shortly before Kore, so it is Jupiter XLVIII. Its orbit ranges from 9,040,000 to 20,018,300 miles from Jupiter, and takes 731.10 days (just over two Earth years). The orbital tilt is 140.1 degrees.

The name is pronounced se lee nee, and commemorates one of Jupiter's daughters, a mountain nymph.

CARME GROUP

[J 54] Eukelade

This name is pronounced you kell ah dee. The moon was a Sheppard discovery in 2003, and is listed as Jupiter XLVII. It has a diameter of 2.5 miles, and an orbital period of 735.2 days. It has an inclination of 165 degrees, with a distance ranging from 10,465,900 to 18,719,600 miles. Eukelade was a Muse and daughter of Zeus.

PASIPHAE GROUP

[J 55] S 2003 J4

Scott Sheppard found this as yet unnamed moon in 2003. It has a diameter of 1.2 miles. It averages 14,637,500 miles from Jupiter with an inclination of 149.2 degrees, and an orbital period of 739.29 days.

[J 56] Pasiphae

Pasiphae, or Jupiter VIII, was discovered in 1908 by Philbert Melotte (1880-1961). Pasiphae is 18 miles in diameter, a giant in its region. Actually, it is believed Pasiphae experienced a collision which chipped most or all the other moons in its group off of Pasiphae. The orbital inclination is 143 degrees, with a period of 746.08 days. Pasiphae's orbit runs from 10,544,800 to 19,381,300 miles.

The Pasiphae of legend was the mother of the Minotaur.

[J 57] Hegemone

Hegemone was another 2003 Sheppard discovery, leading to its being Jupiter XXXIX. With a period of 747.50 days, it averages 14,719,300 miles from Jupiter with an inclination of 151 degrees. The size is estimated to be no more than 2 miles in diameter, with a visual magnitude of 15.9.

Hegemone was a Grace and a daughter of Zeus who served as a goddess of plants. Her name means mastery. The pronunciation is heh geh moh nee.

CARME GROUP

[J 58] Arche

This moon's name is pronounced ar kee, not like Jughead's friend. Recorded as Jupiter XLIII, Scott Sheppard discovered it in 2002. It takes 746.19 days to go around Jupiter at a distance of 12,530,700 to 16,935,900 miles. The orbit is tilted 162 degrees to Jupiter's equator. Arche has a diameter of 1.8 miles. It is named for a Muse, and hence a daughter of Zeus.

[J 59] Isonoe

The name is pronounced eye soh noh ee. Scott Sheppard discovered Jupiter XXVI in 2001. It takes 750.13 days in an orbit with a semimajor axis of 14,780,200 miles and an inclination of 169 degrees. It has a diameter of 2.4 miles. Isonoe was another of Jupiter's love affairs.

[J 60] S 2003 J9

Another Sheppard discovery in 2003, it is a tiny 0.6 mile diameter. It takes 752.84 day orbit at an average of 14,815,000 miles, inclined 165 degrees.

[J 61] S 2003 J5

This 2003 Sheppard discovery orbits Jupiter in 758.34 days at an average distance of 14,888,000 miles with an inclination of 167 degrees. It has a diameter of 2.5 miles.

NO GROUP

[J 62] Sinope

Seth Nicholson discovered this moon on July 14, 1914, and it became Jupiter IX. It orbits from 11,325,600 to 18,748,800 miles from Jupiter in 762.33 days. It is 24 miles in diameter, making it something of a giant among Jupiter's remote satellites. With an inclination of 153.12 degrees there has been some argument that it should be regarded as part of the Pasiphae group, but those moons are all a neutral gray color, while Sinope is distinctly reddish. Its size also suggests an independent source rather than a fragment.

Sinope must have been a beauty, as both Apollo and his father Jupiter desired her. A city on the Black Sea was named for her in ancient times.

PASIPHAE GROUP

[J 63] Sponde

Sponde, or Jupiter XXXVI, was a Sheppard discovery in 2001. It takes 771.6 days to complete an orbit with a semimajor axis of 15,070,572 miles from the planet, ranging from 8,397,800 to 21,742,360 miles. The inclination is 156 degrees. Sponde is 1.2 miles in diameter.

Sponde was a daughter of Jupiter and one of the Horae presiding over the hour when libations are poured after lunch. More proof, if proof was needed, about what fun guys astronomers are.

[J 64] Autonoe

Sheppard's 2002 discovery is also Jupiter XXVIII. With a semimajor axis of 15,068,300 miles it takes 772.17 days for its 150 degree inclined orbit. Autonoe is 2.5 miles in diameter.

Autonoe was one of Jupiter's lovers and mother of the Graces. It is pronounced aw toh noh ee.

[J 65] Megaclite

Megaclite was discovered by Scott Sheppard in 2001, and is Jupiter XIX. It averages 15,331,000 miles from Jupiter, taking 792.444 days to circle the planet. The inclination is 148 degrees and it is 3.3 miles in diameter.

She was another of Jupiter's love affairs.

[J 66] S/2003 J-2

The most distant known of Jupiter's moons has not yet been named. Sheppard found it in 2003, orbiting 15,270,540 to 22,350,760 miles from the planet and taking 981.55 days (2.69 years) for a complete orbit. The tilt is 152 degrees, and it is 1.2 miles in diameter.

SATURN

Saturn could be said to have billions, if not trillions, of satellites, since the rings are composed of hordes of tiny particles. I decided it would be impractical to include that many in a book. Thus here we list moons at least 1000 feet in diameter, with a strong guarantee that there are more to be found. So far, photographs sent back from the Cassini probe and earlier spacecraft visiting Saturn suggest that at least 150 objects smaller than that but clearly not a mere snowflake can be detected within the rings. The rings and any embedded moonlets have orbits that are precisely aligned with the planet's equator, i.e. the orbital inclination is close to or at zero.

[S 1] S 2009 S 1

This moonlet has a length of about 1300 feet and a thickness of nearly 1000 feet. It is 72,600 miles from Saturn, located in the outer B ring, and was discovered on July 26, 2009 from the shadow it cast on the ring at Saturn's equinox (when the Sun is aligned precisely with Saturn's equator). This moonlet takes 0.47 day to orbit the planet. Carolyn Porco headed the group making the discovery.

[S 2] Pan

Mark Showalter discovered Pan on July 16, 1990 from photographs sent back by Voyager 2. It is listed as S XVIII. Pan takes 0.57505 day (13.8 hours) to orbit Saturn at 82,956 miles. The orbit is so nearly circular that the periapsis and apapsis vary from the semimajor axis value by only 2.5 miles each. The orbit is within the 200 mile wide Encke Division of the rings, and it is assumed Pan is responsible for creating the division and keeping it clear. Pan has an equatorial ridge, material picked up from the rings. Pan almost certainly has gaps and vacancies within it, because its density is only 0.42. The albedo is a fairly high 15, suggesting icy deposits on the surface. The escape velocity is only 19 feet per second.

This moon is classified as a ring shepherd, meaning it helps keep the rings in place, and so was named for a god of shepherds and son of Mercury. There is also an asteroid named 4450 Pan.

[S 3] Daphnis

Daphnis was discovered on Cassini probe photographs by a group headed by Carolyn Porco on May 6, 2005. It takes 0.59408 day to orbit, with an inclination of 0.0016 degree. The orbit is within the 26 mile wide Keeler Gap in the rings, 84,770 miles from Saturn, with a variation of 6 miles plus or minus.

Daphnis is 6 miles by 5.5 miles by 4 miles, with the long axis permanently pointed at the planet. The albedo is 5.

Daphnis was a shepherd and descendent of Titans in mythology, with the power of an oracle.

[S 4] Atlas

Atlas, or Saturn XV, was discovered by Richard Terrile (1951-) from photographs sent back by Voyager 2 in 1980. It is 25.5 by 22 by 12 miles, a mishapen lump like many small moons. It orbits 85,493 miles from Saturn, with a deviation of just six miles either way. The orbit takes 0.60169 day, and has an inclination of 0.0032 degree. It shape most resembles a saucer, suggesting material from neighboring rings has fallen on Atlas and reshaped it. The albedo is 4, and the density is 0.46, suggesting it is not solid. It is magnitude 19.0 when Saturn is at opposition for Earth.

Atlas is named for the Titan who held the Earth on his shoulders. His father was Iapetus (S 24).

Prometheus from a distance of 23,000 miles.

[S 5] Prometheus

Saturn XVI was discovered in 1980 from photographs sent back by Voyager 1. It is 84.5 by 49 by 39 miles. It is a shepherd moon for the inner F ring, 86,555 miles from Saturn in an orbit that takes 0.61299 day. The tilt is 0.008 degree. It has an albedo of 6, perhaps related to a surface layer of fine grained material. The surface seems to be very old from the number of craters. The density is only 0.48, suggesting an object only partially compacted. It is in an orbital resonance with Pandora, approaching within 870 miles (among the closest for any two moons) every 6.2 years. Prometheus is magnitude 15.8.

Prometheus is named for the Titan who brought fire to mankind. His father was Iapetus (S 24), and brother Epimetheus (S 7). An asteroid and a volcano on Io (J 5) are also named Prometheus.

[S 6] Pandora

Pandora, or Saturn XVII, was discovered in 1980 on photographs sent back by Voyager 2. It is 65 by 50 by 39.75 miles, with many craters, including two close to 20 miles in diameter. It is 88,008 miles from Saturn, a distance that varies by only six miles either way, shepherding the outer F ring. Its orbital period is 0.62850 day. The inclination is 0.05 degree. The density is only 0.49 and the albedo 6. It is in a

3:2 orbital resonance with Mimas, meaning that every three times Pandora goes around Saturn Mimas (S 10) will complete 2 orbits. Pandora is magnitude 16.4.

Pandora is named for the woman who opened the box that released all the cares and woes into the world following her involvement with Epimetheus (S7).

[S 7] Epimetheus

Epimetheus, or Saturn XI, was discovered in 1977 by John Fountain and Stephen Larson (however, see the discussion on the discovery of Janus, the next moon). Epimetheus is 81 by 71 by 65.8 miles, at a distance of 94,033 or 94,064 miles from Saturn. This moon is called a co-orbital with Janus. They switch places with one another every four years. In the closer orbit it takes 0.69433 day to go around the planet. The density is only 0.64. Epimetheus is magnitude 15.6 at opposition.

This moon has two types of surface, dark and smooth (reminiscent of our Moon's maria) and a brighter, yellowish region with fractures. There are several craters more than 18 miles in diameter, including a very large one near the south pole. This may explain a very diffuse ring of material seemingly originating on Epimetheus. An interesting note is that Epimetheus, like most close moons, has synchronous rotation, meaning that it rotates on its axis in the same time that it takes to go around the planet. But when it swaps places with Janus its orbital period lengthens to 0.69466 day, which must create stresses within the body, as the rotational period is unchanged. There is no evidence of any effect such as produces the volcanoes on Io, but perhaps the fractures seen on part of the surface are evidence of these stresses.

Epimetheus was the nearly forgotten brother of Prometheus (S 5), jointly responsible for creating mankind, and a lover of Pandora (S 6).

[S 8] Janus

Janus, Saturn X, was discovered in 1966 by the French astronomer Audouin Dollfus. The similarity of its orbit with Epimetheus created considerable confusion, until astronomers realized that they were dealing with two different objects. Richard Walker was the first to make this suggestion.

Janus is 126 by 115 by 95 miles, and its distance from Saturn swaps with Epimetheus between 96,064 and 94,033 miles. The inclination is 0.163 degree, with an albedo of 71. There are several craters of close to 20 miles diameter, and some linear features, more than are found on Prometheus, fewer than on Pandora. Janus is magnitude 14.4.

Janus is named for the two faced god who ruled the start of the year, and for whom the month of January is named.

[S 9] Aegaeon

Aegaeon (pronounced eye j eye on) was discovered August 15, 2008 on Cassini spacecraft photos by Carolyn Porco. It is Saturn LIII. It is only a third of mile in diameter, and 104,018 miles from Saturn, with a variation of just four miles either way. The tilt is 0.001 degree. Its orbital period is 0.80812 day. It is in the G ring, and has an orbital resonance of 7:6 with Mimas.

Aegaeon was quite a character in mythology, guarding the gates of Tartaros, the son of Gaia and Uranos, he was a giant with great strength and ferocity. He helped the gods overthrow the Titans. He must have been an interesting sight, with one hundred hands and fifty heads. He was known to the Romans as the sea goat, suggesting some sort of link, unexplained, to Capricorn.

[S 10] Mimas

While some prefer to pronounce this name my mass, its discoverer, William Herschel, and I, prefer mee mas. Herschel discovered this moon in 1789. It is also known as Saturn I, although for being nearest Saturn when the numbers were handed out, not for being the first discovered. It is magnitude 12.8.

Mimas is 258 by 244 by 237 miles. Its orbit ranges from 112,961 to 117,478 miles from Saturn in a period of 0.942422 day at an inclination of 1.566 degrees to Saturn's equator. The density is 1.15, suggesting it is largely composed of ices with a small rocky core. Mimas has a very prominent crater 81 miles across named for Herschel. This crater's walls are three miles high, the floor is six miles deep, and there is a 3.7 miles high central peak which is believed not to be a splash effect like most central peaks, but an actual chunk of the object that impacted to form the crater sticking out of the ground. The impact was so severe that the opposite side of Mimas has fractures.

Mimas, S10, showing the large crater named for William Herschel

The gravitational effect of Mimas causes the Cassini Division in Saturn's rings, as well as the Huyghens Gap. Mimas is in a 2:1 resonance with Tethys and 2:3 with Pandora.

In mythology Mimas was a giant son of Gaia. He had serpents for legs.

[S 11] Methone

This one's pronunciation is a noncontroversial meh thoh nee, otherwise Saturn XXXII. It was discovered in June 2004 from photographs taken by the Cassini spacecraft. The orbit is 120,735 to 120,759 miles from Saturn, in a period of 1.00957 days. The inclination varies from 0.003 to 0.020 degree. The diameter is roughly two miles.

Methone was one of seven beautiful daughters of one of the giants, Alkyonides. The moons Anthe (S 12) and Pallene (S 13) are named for two of her sisters.

[S 12] Anthe

Saturn XLIX, Anthe, was discovered in photos from the Cassini spacecraft in 2004. It ranges from 122,760 to 122,784 miles from Saturn in 1.0365 days. The inclination is 0.1 degree. It is little more than half a mile in diameter.

Anthe's name means flowery. She was one of seven allegedly beautiful daughters of a giant. The moons Methone (S 11) and Pallene (S 13) are named for two of her sisters.

Methone, S11, appears to have the smoothest surface in the Solar System

[S13] Pallene

Pallene (pronounced pah lee nee) is another 2004 Cassini discovery, Saturn XXXIII, but was also found on a Voyager 2 photograph from 1981. It is 4 by 4 by 2.5 miles, with the long axis permanently pointing to Saturn. It is 131,780 to 131,872 miles from the planet, taking 1.15375 days to orbit. The inclination varies from 0.178 to 0.184 degree from gravitational effects by Enceladus. The faint E Ring is created by dust kicked off the surface of Pallene by impacts.

This is another beauteous sister in the trio of Methone (S 11) and Anthe (S 12). Their father, Alkyonides, had an estate also named Pallene. It was near Corinth.

[S 14] Enceladus

Enceladus, or Saturn II, was discovered by William Herschel on August 28, 1789 using a telescope he was able to afford to make as a by product of the fame incurred from his 1781 discovery of the planet Uranus. Enceladus is 318.6 by 312 by 308.6 miles. It is 147,767 miles from Saturn, and orbits in 1.370218 days, or 32.9 hours. The inclination is 0.019 degree. Its density is 1.61, a bit higher than average for Saturn's moons, but still low enough to assure it is mostly ices. The albedo is a very high 90%+. It is the source of the particles in Saturn's E ring. This moon is in an orbital resonance of 2:1 with Dione (S 18). Enceladus is magnitude 11.8 at opposition.

As is true for nearly all triaxial moons, the long axis points to the planet, and the moon experiences synchronous rotation on the shortest axis.

Enceladus, S 14, from a distance of 41,000 miles

Enceladus has a very thin atmosphere (<1/10,000 of Earth's). The composition is 91% water, 4% nitrogen (N2), 3.2% carbon dioxide, and 1.7% methane.

There is heavy cratering in the mid to far northern latitudes, with lighter cratering around the equator. There are numerous craters in the 6 to 12 mile diameter range. The distribution of surface features suggests that there have been several episodes of resurfacing, whether from cryovulcanism or another source is uncertain. There is also deformation caused by viscous effects. Surface fractures are up to 300 yards wide. Collapse pits 150 to 900 yards wide are found near some fractures. Rifts, or canyons, 120 miles long, 3 to 6 miles wide, and half a mile to nearly a mile deep cross the surface. The south polar region has water ice with a greenish tint. The so-called tiger stripes are the location for volcanic plumes of water ice, with a slight mix of propane (C3H8), ethane (C2H6), and acetylene (C2H2). The plumes are ejected into space at a speed of 1360 miles per hour (the escape velocity is 538 mph). The salty/organic nature of the plumes means that Enceladus must have either a salty subsurface ocean, or caverns of salty lakes and seas. The heat which melts the ices and drives the plumes' ejection probably comes from a combination of tidal heating by Saturn and other moons, and radioactive decay.

Craters on Enceladus are named for characters in the *Arabian Nights,* while ridges, plains and grooves are named for places in that book.

In mythology Enceladus was a son of Gaia and Uranus, living in Thrace. His legs were serpeants with dragon scales for feet. After losing in the war between gods and giants he was buried underneath Mount Etna.

Tethys, S 15, showing craters with slumping walls

[S 15] Tethys

Giovanni Cassini discovered Tethys on March 21, 1684 from Paris Observatory (where his family remained in charge until the French Revolution). It is also known as Saturn III. It is 669 by 656 by 654 miles. At 182,958 miles from Saturn it takes 1.887802 days to orbit. The inclination is 1.12 degrees. Tethys has a density of 0.98, just about that of pure water ice.

The leading hemisphere is brighter than the trailing one, which has a reddish tint. There are a number of craters more than 25 miles in diameter, the largest being Odysseus at 300 miles. The floor of this crater is two miles deep, and the rim is raised three miles high. Despite this, most craters show signs of slumping or settling, as the ice is unable to sustain the weight of pronounced features. There are also faults and graben extending for 1200 miles. Tethys seems to be almost pure water ice, possibly with some hematite from Phoebe or Hyperion. Tethys is magnitude 10.2.

Tethys was a sea goddess and wife of Oceanos, mother of all or most river goddesses. Iapetus, Rhea, and Tethys were not named by their discoverer Cassini, but by John Herschel in 1847.

[S 16] Telesto

Saturn XIII was discovered in 1980 by Bradford Smith and Harold Reitsema, and given the name of a daughter of Tethys and Oceanos. It is 20.5 by 15 by 12.5 miles. Shortly after its discovery P. Kenneth Seidelmann realized that it was travelling in exactly the same orbit as Tethys, but sixty degrees ahead of Tethys, at what is called the L4 position (named for the French astronomer who first discovered this was one of five stable locations for shared orbits, Joseph-Louis Lagrange, 1736-1813). In addition to the name

Telesto, it is also called Tethys B. It has relatively few craters, especially small ones. It is magnitude 18.6, adding to making it hard to see near much brighter Saturn.

[S 17] Calypso

Calypso was discovered in 1980 by Dan Pessau and P. Kenneth Seidelmann. It is also known as Saturn XIV and Tethys C. Calypso is 18.6 by 14.3 by 8.3 miles. Calypso is not named for a dance, but for a daughter of Tethys and Oceanos who hosted Odysseus for a year. Its surface is smoothed by impacts with particles from the E Ring.

It is regarded as being in the L5 shared orbit with Tethys (S15), but because the orbit of Calypso is inclined by 1.56 degrees (somewhat more than Tethys or its other Lagrangian moon, Telesto), it actually drifts about the L5 point. The average orbit, though, is 60 degrees behind Tethys, as is proper for a Lagrangian satellite. Calypso is magnitude 18.7.

Calypso's name is shared with asteroid 106 Kalypso.

[S 18] Dione

Dione, or Saturn IV, was discovered on March 21, 1684 by Giovanni Cassini, for whom the Cassini spaceprobe, which itself discovered several moons, is named. Dione is 700 by 697.4 by 685 miles. It is 234,363 miles from Saturn. (Note that this is almost the same distance the Moon is from Earth.) It takes 2.736915 days to orbit the planet. (Do you understand why our Moon takes ten times as long?) The inclination is 0.019 degree, and the albedo is well over 90. The apparent visual magnitude is 10.4, about 75 times too faint for the unaided eye to see. The escape velocity is 1600 feet per second. The density is 1.47, indicating it is probably more than half water ice. An atmosphere of oxygen has been detected, with a density 5×10^{-10} of Earth's atmospheric density. Unlike Earth's oxygen, this does not have a biogenic origin, being the result of solar ultraviolet disrupting water ice molecules.

The surface is scarred by chasms, ridges, icy cliffs, and what may be rilles similar to those found on our Moon (although on Dione they are called fossae). The largest crater, which is located near the south pole, is named Evander, and is about 220 miles across. Most of Dione's craters lack much relief due to slumping.

Dione was the mother of Aphrodite, and probably some of the other gods.

Dione, S 18, from 69,990 miles away, with edge-on rings and two small moons in the background

[S 19] Helene

Just as Telesto is a satellite sharing Tethys' orbit, Helene shares the orbit of Dione, traveling at the L4 position 60 degrees ahead of Dione. Helene was discovered in 1980 by Pierre Laques and Jean Lacacheux. It is 26.7 by 23.6 by 16.7 miles, and has an orbit inclined 0.199 degree. Helene is magnitude 18.4 at opposition.

It is named for Helen of Troy, and is also known as Saturn XII and Dione B. The pronunciation is heh lee nee.

[S 20] Polydeuces

Polydeuces was a 2004 discovery by the Cassini spacecraft. It is 1.8 by 1.2 by 0.6 miles. While it is technically in the L5 position of Dione's orbit, trailing Dione by 60 degrees, because its orbit is inclined more than Dione's (or Helene's) it wanders from 33.9 to 91.4 degrees from Dione, and its orbit can vary by 4757 miles from the normal 234,363 miles of Dione and Helene. However, its orbital period is the same as those two moons.

The name Polydeuces is an alternate form of Pollux (as in Gemini), and has nothing to do with holding too many cards of value between an ace and a trey in a poker hand. It is also called Saturn XXXIV and Dione C.

[S 21] Rhea

Rhea, or Saturn V, was discovered December 23, 1672 by Giovanni Cassini. It is 950 by 948 by 947 miles. It orbits 327,334 miles from Saturn in a period of 4.518212 days. The orbital tilt is 0.345 degree. It has a density of 1.236. The surface gravity is 2.7% of the surface gravity on Earth. It is believed to be about 25% rocks, with the remainder ices. Rhea is magnitude 9.6.

Rhea, S 21, from 9500 miles away

Rhea's surface shows heavy cratering with fractures in the surface. Two large craters are 270 and 320 miles in diameter. There is a 31 mile diameter crater from which bright rays radiate, similar to rays found coming from some lunar craters. This is interpreted to mean the crater is likely very young, perhaps less than 100 million years. The trailing hemisphere shows fewer craters but more fractures. Rhea has a very thin atmosphere (10^{-12} Earth's density) that is 72% oxygen and 28% carbon dioxide.

Rhea is named for a daughter of Ouranos and Gaia, and was herself the mother of many of the Greek gods, including Zeus.

[S 22] Titan

Titan, or Saturn VI, was discovered March 25, 1655 by Christiaan Huyghens. Its orbit ranges from 736,928 to 788,654 miles from Saturn in 15.945 days. The orbit is tilted 0.34859 degree. Its density is 1.88. Titan is in a 3:4 orbital resonance with Hyperion. Titan is magnitude 8.4, a bit too faint for unaided eyes, but within reach of good binoculars. The diameter is 3200 miles, and it has an albedo of 22, due almost entirely to clouds in its atmosphere. The surface gravity is 14% that of Earth, and the escape velocity needed to launch from Titan is 1.64 miles per second, not far from that needed to launch from our Moon.

Titan has been examined by the following spacecraft, all from the United States: Pioneer 11 in 1979; Voyager 1 in 1980; Voyager 2 in 1981; Cassini starting in 2004; landed upon by Huyghens on January 14, 2005. This was only the second moon to be landed upon, after Earth's Moon.

Titan has an atmosphere 45% denser than Earth's atmosphere, although the components are different. Both have a majority of nitrogen (N_2 molecules), to which Titan adds 4.9% methane (CH_4), hydrogen (H_2, 0.2%), and trace amounts of ethane, diacetylene (C_4H_2), methylacetylene (CH_3CCH), acetylene (C_2H_2), propane (C_3H_8), cyanoacetylene (C_3HN), hydrogen cyanide (HCN), carbon dioxide, carbon monoxide, cyanogen ($[CN]_2$), argon, and helium. Some of the hydrocarbons and the carbon dioxide will precipate to the surface as snow or rain. Methane breaks down from exposure to solar ultraviolet. Titan has too much methane surviving such exposure in its atmosphere for it to be simply there from outgassing on the surface. The origin of this methane is unknown, but is either from some very exotic chemistry, or the product of some very exotic life forms.

The surface temperature is about –290F, a bit too chilly for its distance from the Sun, so the clouds in Titan's atmosphere, which also hinder observations, must be cooling it a little.

Polar regions of Titan show lakes up to 300 miles across, although they seem to be fairly shallow, mostly less than 20 feet deep. The lakes seem to occur mostly at the pole experiencing winter. The summer pole shows empty basins. At least one equatorial region lake has been observed. The contents of the lakes is not water, but liquid methane. Titan has mountains up to a mile high, and some craters, the largest 275 miles across. There are some cryovolcanoes powered by tidal flexing resulting from Titan's mildly elliptical orbit. These may be drawing material from a layer of liquid water created by tidal friction. Such a layer would probably be several dozen miles in thickness, and a couple dozen miles below the surface. "Sand" dunes are composed of grainy particles of ice and ammonium sulfate.

Titan was named for the race of giants that fought the Olympian gods of Greece.

[S 23] Hyperion

Hyperion, Saturn VII, was discovered on September 16, 1848 by William Bond (1789-1859) and his son George (1829-1865). William Lassell independently discovered it a few days later. The Bonds were the first Americans to discover a planetary satellite. Hyperion is one of the odder moons. It is 223.5 by

165.2 by 127.5 miles, and is shaped somewhat like an arrowhead. It is the largest moon to deviate in a major way from being spherical. It is 919,707 miles from Saturn, taking 21.27661 days to orbit. The orbit is inclined 0.43 degree. Unlike the closer Saturnian moons we have listed [1] through [22], Hyperion is not gravity locked, thus not keeping the same face always to the planet. Its rotation is chaotic, with the axis of rotation drifting unpredictably. It is magnitude 14.4 on average at opposition.

Hyperion has the remarkably low density of 0.544 (far less than water), suggesting it has large cavities within its body. The estimate is that it is 40% hollow. The albedo is 25, and it seems to be covered with a reddish material similar to that found on Iapetus. The largest crater is over 75 miles across and more than six miles deep. Craters on Hyperion generally have deep, sharp edges, with dark material at the bottom.

Hyperion in mythology was the lord of light and in charge of watchfulness. He was the father of Helios and Selene (the Sun and Moon), their mother being Thea.

Hyperion, S 23, from 48,000 miles, showing its odd shape and deep, dark craters

[S 24] Iapetus

This name is variously pronounced yah pet us or eye ap et us. In older texts it sometimes appears as Japetus. Giovanni Cassini discovered this moon on October 25, 1671. It is 2,211,269 miles from Saturn in an orbit inclined 15.47 degrees. It takes 79.3215 days for an orbit. The hemispheres are radically different in coloration, east having an albedo of 5, while the western hemisphere has an albedo of 52. The surface gravity is 2.3% that of Earth, with an escape velocity of 1879 feet per second. The density is 1.088, suggesting it is 20% rocky, with the remainder ices. Iapetus is 926.5 by 926.5 by 884.3 miles.

The two faces of Iapetus, showing the great difference between dark and light sides

With the odd range of albedos Iapetus is magnitude 10.2 when west of the planet, and 11.9 when to the east. It is heavily cratered, with the largest being 360 miles in diameter. The walls of this crater are very steep, and it is as much as ten miles deep. There are five other craters that are more than 210 miles across. The dark hemisphere has been named Cassini Regio in honor of the moon's discoverer. The dark material is believed to be organic compounds, particularly polymers of hydrogen cyanide (HCN). The material is thin on the ground, as shown by subsequent cratering, in many places no more than a foot thick. At one time it was assumed the dark material had splashed onto the surface from an impact on Phoebe, but today the belief is that this material migrated due to solar heating from the western hemisphere, being residue from the sublimation of water ice. There is a ridge running along the equator for 800 miles in the dark hemisphere. This ridge runs up to 12.5 miles wide, and has many hills up to 8 miles high, with some peaks reaching more than 12 miles high. This is believed to have been caused by an impact on Iapetus that left material in a low orbit around the moon, within its Roche limit. The material ultimately fell back to the surface, strongly concentrated along the equator.

From Iapetus Saturn would be about four times the size the Moon is to Earthlings, 2 degrees across. This is the closest moon with a good view of the rings, since the others are so nearly in the same plane as the rings.

Iapetus was the Titan of life, and the father of Prometheus, Atlas, and others.

[S 25] Kiviuq

Brett Gladman and John Kavelaar discovered this moon in 2000, and on the suggestion of a Canadian astronomer gave it a name taken from Inuit mythology rather than Greco-Roman. Otherwise it is Saturn XXIV. Kiviuq has a diameter of ten miles. Its distance from Saturn ranges from 4,707,860 to 9,320,337 miles in an orbit that takes 448.16 days. The orbit is inclined 45.71 degrees to Saturn's equator, and hence to the plane of the rings, assuring a good, if distant look at them. This small moon rotates in 21 hours 49 minutes, the first of Saturn's moons other than Hyperion that we have seen that is not gravity locked on the planet. It has a light reddish color and may be a surviving fragment along with the moons Siarnaq and Paaliaq of the break up of a larger body.

The Inuit, the native people of much of northern Canada, had a number of myths about Kiviuq, who was supposed to be a legendary hero of the Inuit, possessing some of the qualities of Paul Bunyan and Odysseus.

[S 26] Ijiraq

Saturn XXVI was also discovered in 2000 by Brett Gladman and John Kavelaar. It is seven miles in diameter. Its distance from Saturn ranges from 5,825,730 to 9,286,656 miles, taking 451.77 days to orbit. The orbital tilt is 50.212 degrees. It is much redder than other moons of Saturn, and lacks the water ice found to be such an important component of most other moons.

Ijiraq was a shapeshifting deity of the Inuit.

[S 27] Phoebe

Phoebe was discovered by William Pickering on March 17, 1899. It was the first moon ever discovered using photography rather than looking through a telescope with eyes. It is Saturn IX. Phoebe is 136 by 136 by 126.7 miles. The orbit is inclined 151.78 degrees, meaning it is a retrograde orbit. Its elliptical orbit is 6,747,730 to 9,246,735 miles from Saturn, and takes 545.09 days to orbit. The rotational period is 9 hours 16 minutes 55.2 seconds (not gravity locked). The density is 1.638, relatively high for Saturn's moons, and the albedo is 6, very low for Saturn's moons. It is magnitude 15.8, agreeing with the albedo that the surface is dark.

Phoebe shows heavy cratering, with some up to 50 miles diameter. One has walls 10 miles high. The surface has ice beneath a blanket of dark material which ranges from 980 to 1600 feet thick. There are patches of carbon dioxide ice on the surface, but Phoebe appears to be 50% rocks. Phoebe is apparently differentiated, with a rocky core. It is believed to have formed in the Kuiper Belt and been captured into Saturn orbit during an early period of the Solar System. Radioactives would have warmed it, permitting the differentiation. There is a very slight ring of material knocked off this moon sharing its orbit.

Phoebe appears to have a thin coating of dark material overlying an icy body

Voyager 2 photographed Phoebe from a distance of 1.75 million miles in September 1981. The Cassini spacecraft was deliberately targetted to fly near Phoebe on its way in towards Saturn, and photographed it from only 1285 miles on June 11, 2004.

Phoebe is named for a Titan who was related to the Moon.

[S 28] Paaliaq

Brett Gladman and John Kavelaar discovered Saturn XX in October 2000. It is fourteen miles in diameter. Paaliaq has an elliptical orbit taking it 5,977,486 to 12,793,038 miles from Saturn, with an inclination of 46.51 degrees. It takes 692.98 days to go around Saturn. It has a light red color, similar to the moons Kiviuq and Siarnaq, suggesting a common origin. It rotates in nineteen hours.

This moon was named for a fictional shaman in a novel.

[S 29] Skathi

Gladman and Kavelaar found Saturn XXVII in 2000. Its 149.089 degree orbital inclination indicates a retrograde orbit, 7,343,125 to 12,134,660 miles from the planet. It is only 4 miles in diameter, and takes 732.52 days (2 years, 2.52 days) for an orbit.

Skathi was a giantess, married to one of the Vanir, Niord, in Norse legend.

[S 30] Albiorix

Matthew M. Holman discovered this twenty mile diameter moon in 2000. It is also Saturn XXVI. Its 5,786,196 to 15,887,816 mile elliptical orbit takes 774.58 days to go around Saturn. The orbit's tilt is 38.042 degrees. Albiorix rotates in 13 hours 19 minutes. Its spectrum resembles that of the much smaller moons Tarvos (#41 below) and Erriapus (#33 below), which may be fragments knocked off Albiorix.

Albiorix was a giant in Gallic mythology similar to the Roman god Mars in personality and role.

[S 31] S 2007 S2

This as yet unnamed moon was discovered on May 1, 2007 by Scott Shepperd, David Jewett, Jan Kayna and Brian Marsden. It has a diameter of four miles. The 10,283,760 mile mean orbit takes 792.96 days, and is inclined 176.68 degrees.

[S 32] Bebhionn

Bebhionn (pronounced bev ee on) was discovered in 2004 by Scott Shepperd, David Jewett, Jan Kayna and Brian Marsden. It is Saturn XXXVII. Bebhionn has a diameter of four miles, and averages 10,652,335 miles from Saturn, taking 838.77 days for an orbit. The orbit is tilted 18 degrees. It rotates in about sixteen hours.

Bebhionn is named for an Irish goddess who ruled over childbirth and was noted for her beauty.

[S 33] Erriapus

Erriapus was discovered in 2000 by Brett Gladman and John Kavelaar. It is Saturn XXVIII. Erriapus has a diameter of six miles, and takes 844.89 days to go around Saturn at distances of 5,651,118 to 15,770,890 miles. The orbit is tilted 38.109 degrees. The spectrum matches that of Albiorix [S 30] and Tarvos [S 41].

Erriapus is named for a giant in Gaulish mythology.

[S 34] Skoll

Skoll, Saturn XLVII, was discovered June 26, 2006, by Scott Shepperd, David Jewett, Jan Klayna, and Brian Marsden. It ranges from 6,315,416 to 17,166,666 miles from the planet in an orbit that takes 862.37 days. The orbit is tilted 155.624 degrees, thus moving retrograde. Skoll is 4 miles in diameter.

In Norse mythology Skoll, whose name means treachery in Old Norse, was a giant wolf, son of the Fenris monster wolf. Skoll chased the chariot carrying the Sun across the sky, and at Ragnarok will catch and eat it, a hotter meal than India could ever provide. Hati [S 48] is named for his brother.

[S 35] Siarnaq

Siarnaq, Saturn XXIX, was discovered in 2000 by Brett Gladman and John Kavelaars. With a diameter of 25 miles it is a relative giant among Saturn's outer moons. It has a mean distance from the planet of 11,039,268 miles, taking 884.89 days for an orbit. The inclination is 45.798 degrees.

Siarnaq is light red with a spectrum similar to Kiviuq and Paaliaq, suggesting they are the shattered results of a collision. It rotates in 6.7 hours. There is a resonance between the periapsis of this moon and Saturn's perihelion.

Siarnaq was a giant in the mythology of the Inuits of Canada.

[S 36] Tarqeq

The team of Shepherd, Jewitt, Kayna and Marsden discovered Tarqeq, Saturn LII, on April 13, 2007. Its orbit's semimajor axis is 11,122,962 miles, taking 894.86 days. The inclination is 49.77 degrees. Tarqeq is thought to be 4.3 miles in diameter.

This moon is named for the Inuit's moon god (which disproves the title of the book *The Moon is Always Female*).

[S 37] S/2004 S-13

This so far unnamed moon is another discovery by Scott Shepherd, David Jewitt, Jan Kayna and Brian Marsden, on May 4, 2005. It averages 11,212,962 miles from Saturn, taking 905.85 days for an orbit. The orbit is inclined 143 degrees.

[S 38] Greip

This moon's name is pronounced like the English word grape. It is Saturn LI, and was discovered on June 26, 2006 by the team of Shepherd, Jewitt, Kayna and Marsden. Its elliptical orbit ranges from 7,033,105 to 15,418,845 miles from Saturn, and takes 906.56 days to orbit it. The inclination is 159.2 degrees. The diameter is about four miles.

Greip was a Norse giantess, and a mother of the god Heimdall, who claimed to have nine mothers.

[S 39] Hyrrokkin

Hyrrokkin (heer rohk kin) was discovered June 6, 1006 by Shepherd, Jewiitt, Kayna and Marsden, becoming Saturn XLIV. It is about five miles in diameter, and at 7,220,944 to 15,358,618 miles from Saturn, takes 914.29 days to orbit the planet. The inclination is 154.3 degrees.

The name in Old Norse means "fire smoked", and is the name of a giantess of great strength, who launched the funeral boat of Baldur.

[S 40] Jarnsaxa

Saturn L's name is pronounced yarn sayksa. It was discovered by the team of Shepherd, Jewitt, Kayna and Marsden on June 26, 2006. Jarnsaxa is 3.5 miles in diameter, and at 9,319,562 to 13,742,952 miles from Saturn has an orbit taking 943.78 days. The orbit is inclined 164.1 degrees.

In Old Norse Jarnsaxa means "iron axe". She was one of Thor's lovers, and one of Heimdall's nine mothers.

[S 41] Tarvos

Tarvos, or Saturn XXI, was discovered by Brett Gladman and John Kavelaars on September 23, 2000. It is about 9.5 miles in diamterer. Tarvos has the second most elliptical orbit of all of Saturn's known moons, ranging from 5,412,160 to 17,642,836 miles from the planet. The full orbit takes 944.23 days. This orbit is inclined 33.825 degrees. Tarvos is a light red, and is suspected of being a fragment of Albiorix [S30] knocked loose in a collision.

The name is from a Gaullish god who, in the shape of a bull, carried three cranes on his back. The full name of the god (not the moon) was Tarvos Trigaranos, and he is shown on a monument on the Seine near Paris, along with Jupiter, Vulcan, and a Celtic war god, Esus.

[S 42] Mundilfari

This discovery in 2000 by Gladman and Kavelaars is Saturn XXV. It has a diameter of just 3.5 miles. and takes 956.7 days to complete an orbit averaging 11,628,720 miles from Saturn. The inclination is 150 degrees.

Pronounced mun dell verr ee, this is named for a Norse god whose name means "moving according to time". He was the father of the Sun and the Moon.

[S 43] S/2006 S-1

Another as yet unnamed moon discovered by the team of Shepperd, Jewett, Kayna and Marsden on June 26, 2006. It has a four mile diameter, and at an average of 11,755,650 miles, orbits in 972.45 days. The inclination is 175.4 degrees.

[S 44] S/2004 S-17

This still nameless discovery by Shepperd, Jewett, Kayna and Marsden was discovered May 4, 2005 on photgraphs made the previous year. It has a diameter of just 2.5 miles. Its semimajor axis is 11,860,600 miles, taking 985.45 days for an orbit that is inclined 162 degrees.

[S 45] Bergelmir

This one is alternatively pronounced as either bear jel meer or bair yel meer. Whichever you choose, it was discovered May 4, 2005 by the team of Shepperd, Jewett, Kayna and Marsden. Bergelmir is also Saturn XXXVIII. It averages 11,863,584 miles from the planet, taking 985.83 days for an orbit inclined 134 degrees. It has a diameter of four miles.

Bergelmir is named for a Norse giant, a grandson of Ymir.

[S 46] Narvi

Another discovery by the team of Shepperd, Jewitt, Kayna and Marsden, Narvi is Saturn XXXI. It is believed to be 4.1 miles in diameter, and runs from 8,095,480 to 15,908,874 miles from the planet, taking

1008.45 days to orbit. The inclination is 109 degrees, remarkably close to a polar orbit. It was discovered in 2003.

Narvi was a son of Loki who was killed by his twin brother when the brother was turned into a wolf by Norse gods trying to get back at Loki for his tricks.

[S 47] Suttungr

Suttungr was discovered by Brett Gladman and John Kavelaars in 2000. It is about four miles in diameter, and has a semimajor axis 12,158,560 miles from Saturn. Also called Saturn XXIII, it takes 1022.82 days to orbit. The inclination is 151 degrees. It may be debris knocked off Phoebe by impacts, given the similarity in spectra.

Suttungr was a Norse giant who was given mead which when drunk inspired poetry. Suttungr hid this mead in a mountain, where Odin found and stole it.

[S 48] Hati

Saturn XLIII has a diameter of 3.5 miles. It was discovered by Shepherd, Jewitt, Kayna and Marsden on May 4, 2005. Hati's orbit is 8,683,578 to 15,811,340 miles from Saturn, taking 1033.05 days to go around it. The inclination is 165 degrees.

Hati's name means "he who hates". He was a twin brother of Skoll [S 34], and chased the Moon with the intention of eating it. At Ragnorok Hati was expected to succeed in this meal.

[S 49] S/2004 S-12

This still unnamed moon was discovered by Shepherd, Jewitt, Kayna and Marsden on May 4, 2005. It is believed to be just three miles in diameter. At 12,366,600 miles from Saturn it takes 1048.54 days to orbit. The inclination is 162 degrees.

[S 50] Farbauti

Saturn XL was a 2002 discovery of Shepherd, Jewitt, Kayna and Marsden. It is about three miles in diameter. The semimajor axis is 12,410,560 miles, taking 1054.78 days to orbit. The tilt is 151 degrees.

Farbauti was a storm giant, and father of Loki. The name means "cruel striker", and may be a kenning for lightning.

[S 51] Thrymr

Gladman and Kavelaars discovered Saturn XXX in 2000. It is roughly 3.5 miles in diameter, and 12,592,700 miles from Saturn in a 1078.09 day orbit. The inclination is 151 degrees.

Thrymr was the king of the frost giants in Norse mythology, and was killed by Thor for having stolen Thor's hammer.

[S 52] Aegir

Aegir, or Saturn XXXVI, was discovered by the team of Shepherd, Jewitt, Kayne and Marsden on May 4, 2005. It is about 3.5 miles in diameter. Aegir averages 12,218,880 miles from Saturn, taking 1094.46 days (about thirteen hours short of three Earth years) to orbit the planet. The inclination is 167 degrees.

Pronounced eye gur, this moon is named for the Norse god of the ocean, and king of all sea creatures. He was responsible for throwing great parties for the Norse gods, but seems to have been older than them or the giants. He had nine children, personifications of waves of the sea. His father was Fornjot [S 62].

[S 53] S/2007 S-3

This still nameless moon was discovered April 19, 2007 by Shepherd, Jewitt, Kayna and Marsden. It has a three mile diameter, and orbits an average of 12,742,000 miles from Saturn in 1100 days. The orbit is inclined 177 degrees, meaning it is almost an equatorial orbit, but going in the reverse direction (retrograde, or clockwise) from most Solar System traffic.

[S 54] Bestla

Shepherd, Jewitt, Kayna and Marsden discovered this 4.5 miles diameter moon on March 5, 2005. It is also known as Saturn XXXIX. Bestla's exceptionally elliptical orbit ranges from 2,939,906 to 22,624,490 miles from the planet, taking 1101.45 days to orbit. The inclination is 151 degrees.

For the Norse Bestla was a frost giant, and the mother of Odin.

[S 55] S/2004 S-7

Another discovery, on December 12, 2004, by Shepherd, Jewitt, Kayna and Marsden awaiting a name. It is 3.5 miles in diameter, and ranges from 6,771,130 to 18,785,130 miles from Saturn in an orbit that takes 1102 days. The inclination is 166 degrees.

[S 56] S/2006 S-3

This 3.5 miles diameter moon was discovered by Shepherd, Jewitt, Kayna and Marsden on June 26, 2006. its semimajor axis is 13,088,385 miles, and it takes 1142.37 days to orbit. The inclination is 128.8 degrees.

[S 57] Fenrir

Fenrir, Saturn XLI, was discovered by Shepherd, Jewitt, Kayna and Marsden It is only about two and a half miles in diameter. The orbit is 13,618,930 miles from Saturn, requiring 1260.35 days to complete. The inclination is 143 degrees.

Fenrir has an apparent magnitude of +25, making it one of the fainter known satellites in the Solar System.

In Norse mythology Fenrir was a ferocious wolf, the son of Loki who was bound until Ragnarok when he would create devastation and havoc.

[S 58] Surtur

This 3.5 miles diameter moon is yet another discovery, on June 26, 2006, by Shepherd, Jewitt, Kayna and Marsden. It is Saturn XLVIII. The orbit ranges from a periapsis of 8,747,775 miles to apapsis of 18,935,052 miles, with an inclination of 148.9 degrees. The orbital period is 1242.36 days.

Surtur is a fire giant whose name means the swarthy one. At Ragnarok his duty was to guard the entrance to Muspellheim, domain of the fire giants, and to defeat Frey. There is a volcano on Io (Jupiter) named for him, and he also appears as a character in the science fantasy novel *The Incomplete Enchanter,* by Fletcher Pratt and L. Sprague de Camp.

[S 59] Kari

This 4 miles diameter moon was also discovered on June 26, 2006 by the team of Shepherd, Jewitt, Kayna and Marsden. It is Saturn XLV. The orbit ranges from 9,146,636 to 18,581,294 miles from the planet, taking 1245.06 days. The inclination is 151.5 degrees. It rotates in 7 hours 42 minutes.

Kari is a personification of the wind in Old Norse.

[S 60] Ymir

Ymir (ee mur) is Saturn XIX, discovered in 2000 by Brett Gladman and John Kavelaars. It is eleven miles in diameter, the largest of Saturn's very distant moons. The orbit ranges from 9,264,060 to 18,593,690 miles, with a period of 1315.14 days (3 years 235.14 days). The inclination is 173.125 degrees, meaning it is almost aligned with Saturn's equator, despite having retrograde motion. The apparent magnitude is 21.7, and the albedo 6. The escape velocity is an unchallenging 19.25 miles per hour.

Ymir was the ancestor of all frost giants.

[S 61] Loge

Loge is 3.5 mile diameter moon discovered June 26, 2006 by Scott Shepherd, David Jewitt, Jan Kayna and Brian Marsden. It is Saturn XLVI. Its orbit takes 1300.95 days as it goes from 11,289,283 to 16,257,243 miles from Saturn. The inclination is 165 degrees.

Loge was a fire giant, and not to be confused with Loki, whom he bested in an eating contest.

[S 62] Fornjot

Fornjot (forn yot) is a May 5, 2005 discovery by Shepherd, Jewitt, Kayna and Marsden. It is 3.5 miles in diameter. It is Saturn XLII. Fornjot's orbit ranges from 12,387,069 to 18,097,789 miles, the most distant of any known moons for Saturn. This remote orbit takes 1432.16 days (4 years 1.16 days).

Fornjot was a giant and King of Finland, the father of Aegir [S 52], Kari [S 59], and Loge [S 61].

URANUS

Uranus was the first planet to be discovered, by William Herschel in 1781. He wanted to name it for the King of England, George III, a proposal that was not well received by anyone other than George. The name finally chosen was for the father of Saturn and god of the heavens. Herschel went on to discover the first couple moons and here his choice to use names from Shakespeare was accepted. Today all moons of Uranus are named either from Shakespeare or from the works of Alexander Pope.

You might notice that the Uranian moon system resembles that of Jupiter, in that each has several small close satellites, followed by several large ones, and then at a remote distance a batch of small ones in very elliptical retrograde orbits.

[U 1] Cordelia

Cordelia, or Uranus VI, is 32 by 22 by 22 miles, with the long axis pointed to the planet. Its orbit is 30,895 miles from Uranus, taking 0.335 day to go around. The inclination is 0.085 degree to the equator of Uranus (which, remember, is itself tilted 97.9 degrees). It was discovered January 20, 1986 by Richard Terrile examining Voyager 2 photographs. The albedo is 8, with visual magnitude 23.6 at opposition. Rotation is synchronous with its orbital motion. Cordelia is a shepherd moon for the Epsilon Ring of Uranus.

Cordelia was the youngest (loyal) daughter in Shakespeare's "King Lear".

[U 2] Ophelia

Uranus VII has the same discovery date and circumstances as Cordelia. It is 33.5 by 23.6 by 23.6 miles, with the long axis pointed to the planet. It is 33,387 miles from Uranus, taking 0.3764 day to go around. The inclination is 0.1036 degree. The rotation is synchronous with the orbit, and the albedo is 8. Ophelia is magnitude 23.3. The escape velocity is 40 miles per hour. It is a shepherd of the Epsilon Ring.

Ophelia was the daughter of Polonius in Shakespeare's "Hamlet", and possibly the title character's girl friend. (I had a much more provocative theory in a paper for undergraduate English, which got an A- and "good grief" from the teacher.)

[U 3] Bianca

Bianca is also Uranus VIII. It was discovered January 23, 1986 by Bradford Smith on Voyager 2 photographs. Bianca is 40 by 28.5 by 28.5 miles, with the long axis pointed to Uranus. It is 36,741 miles from the planet, taking 0.4346 day for an orbit. The inclination is 0.193 degree. Bianca has synchronous rotation, and a surface gravity of 0.0008G. It is magnitude 22.5.

Bianca was Kate's sister in Shakespeare's "Taming of the Shrew".

[U 4] Cressida

Cressida, or Uranus IX, was discovered January 9, 1986 by Stephen Synnott on a Voyager photograph. It is 57 by 46 by 46 miles. Its orbit is 38,357 miles from the planet, taking 0.46357 day to go around. The inclination is 0.006 degree to the planet's equator. Cressida has an albedo of 8, and is gray. Rotation is synchronous. It is magnitude 21.6.

Cressida was the tragic Trojan wife in Shakespeare's "Troilus and Cressida".

[U 5] Desdemona

Desdemona, or Uranus X, was discovered by Stephen Synnott on January 13, 1986. It is 56 by 33.5 by 33.5 miles. It is 38,910 miles from Uranus taking 0.47365 day to orbit. The inclination to the planet's equator is 0.11125 degree. Desdemona's albedo is 8, with a gray color. Rotation is synchronous with the long axis pointed to the planet. There is some chance that Desdemona will collide with either Cressida or Juliet at a remote date. Desdemona is magnitude 22.0 at opposition.

Desdemona was Othello's wife.

[U 6] Juliet

Juliet is Uranus XI. It was discovered January 3, 1986 from Voyager 2 photographs by Stephen Synnott. Juliet is 93 by 46 by 46 miles, the most elongated of Uranus's moons. As usual, the long axis points to the planet. It is 39,968 miles from Uranus, taking 0.493065 day to orbit. Inclination to the equator of Uranus is 0.06546 degree. Its albedo of 8 goes with a gray color. It is magnitude 21.1.

If you can't figure out who Juliet is in Shakespeare, why are you reading astronomy instead of learning some literature?

[U 7] Portia

Portia is Uranus XII. Stephen Synnott discovered it on a Voyager photograph on January 3, 1986. It is 97 by 78 by 78 miles. It orbits 41,460 miles from Uranus in 0.513196 day with an inclination to the planet's equator of 0.059 degree. Water ice has been detected on its surface. Portia is magnitude 20.4.

Portia appears in Shakespeare's "Merchant of Venice".

[U 8] Rosalind

Stephen Synnott discovered Uranus XIII on January 13, 1986. It is 43,425 miles from Uranus, taking 0.55846 day to orbit. The inclination is 0.278 degree. It is roughly spherical with a 45 mile diameter. The albedo is 8, with a gray toned surface. It is magnitude 21.8.

Rosalind is named for the Duke's daughter in "As You Like It".

[U 9] Cupid

Mark Showalter and Jack Lissauer found Cupid, or Uranus XXVII, on old Voyager 2 photographs on April 25, 2003. It has an eleven mile diameter, making it the smallest of Uranus's inner satellites. Cupid orbits 46,450 miles from the planet in 0.618 day. The inclination is 0.1 degree. It has an apparent magnitude when Uranus is at opposition of +14.8.

Cupid is named for a character in "Timon of Athens".

[U 10] Belinda

Stephen Synnott found Uranus XIV on January 13, 1986 on photos sent back by Voyager 2. Belinda is magnitude 21.5. It is 79 by 40 by 40 miles, with the long axis pointed at the planet for synchronous rotation. The orbit is 46,735 miles from the planet with an inclination of 0.031 degree. The orbital period is 0.62352 day. Belinda and Cupid can on rare occasions approach within a couple hundred miles of one another.

Belinda is the main character in Alexander Pope's "The Rape of the Lock".

[U 11] Perdita

Perdita, Uranus XXV, was discovered on old Voyager photographs by Erich Karkoschaka May 19, 1999. It is nineteen miles in diameter. Perdita orbits 47,460 miles from Uranus in 0.638 day. The orbital tilt is zero. Perdita is in a 43:44 orbital resonance with Belinda.

Perdita is named for the daughter of Leontes in "A Winter's Tale".

[U 12] Puck

Puck was discovered by Stephen Synnott on December 30, 1985. It has a diameter of 100 miles. Puck orbits 53,408 miles from Uranus in 0.761833 day. The orbital tilt is 0.3192 degree. Voyager 2 managed to get some pictures of Puck, it having been detected early enough to modify the spacecraft's orbit for this purpose. Puck is heavily cratered, the largest being 28 miles diameter. Water ice was found, and the albedo is 11, with a visual magnitude of 19.7 at opposition.

Puck is a sprite appearing in "Midsummer Night's Dream", but in English folklore Puck is the name of a species of mischievous fairies. Three craters on Puck have been named for other such fairy species, including the Irish Bogle.

[U 13] Mab

Mark Showalter and Jack Lissauer discovered Uranus XXVI on August 25, 2003 on old Voyager photographs. Mab has a fifteen mile diameter. Its orbit is 60,700 miles from Uranus, taking 0.923 day to go around. The inclination is 0.1335 degree. It seems to be the source of the Mu ring of Uranus, as material knocked off Mab by impacts forms the ring. The apparent magnitude of this moon is +26.

Mab is a fairy queen mentioned in "Romeo and Juliet".

[U 14] Miranda

Miranda was discovered February 16, 1948 by Gerard P. Kuiper (1905-1973), and is also known as Uranus V. Miranda is 300 by 291 by 290 miles. It orbits Uranus at a distance of 80,295 miles in 1.413479 days. The orbit is tilted 4.232 degrees to the equator of Uranus.

Miranda, U14, showing unusual land forms suggesting this moon broke apart and recombined

Miranda has water ice on its surface, which also includes silicate materials and carbon compounds. This gives it a density of 1.2. There are scarps three miles high, a rectangular feature known informally as the racetrack, craters, and cracks suggesting this moon was at one time broken into pieces which coalesced to form the object we know today. Rotation is synchronous, and surface gravity is 0.008G. It is magnitude 15.8.

Miranda is named for Prospero's daughter in "The Tempest".

[U 15] Ariel

Ariel, or Uranus I, was discovered October 24, 1851 by William Lassell (1799-1880). It is 722 by 718 by 718 miles, and is 118,623 miles from the planet. The orbit takes 2.520378 days. Inclination is 0.26 degree.

Ariel has an albedo of 53, making it by far the brightest moon of Uranus, with an apparent magnitude of 13.7. It seems to be differentiated, with a rocky core having half the mass, and an ice crust for the other half, giving a density of 1.66. The surface shows craters, the largest 48.5 miles diameter. However, only Voyager 2 has ever visited Uranus, and at the time in 1986 the northern half of the planet and all the moons was dark, experiencing winter, so the amount of any moon mapped ranges from 20 to 40 percent. The surface of Ariel that was mapped has water ice and CO_2 ice, with the carbon dioxide ice concentrated towards the trailing hemisphere, which is redder, and the H_2O ice on the leading hemisphere. Voyager 2 got within 79,000 miles of Ariel, mapping 40% to a detail of 2 miles.

The name unites Shakespeare and Alexander Pope. It is the name of a sylph for Pope, and the name of Prospero's spirit servant in "The Tempest".

Ariel, U 15, from 80,000 miles

[U 16] Umbriel

William Lassell also discovered this moon, Uranus II, on October 24, 1851. Umbriel/Uranus II is 726.2 miles in diameter. It is 159,162 miles from Uranus, taking 4.144177 days to orbit. The inclination is 0.205 degree.

Umbriel is the darkest of Uranus's large moons, with an albedo of 26, and magnitude 14.5. The density is 1.39, suggesting an ice/rock ratio of 60/40. It is second most heavily cratered of the moons of Uranus, with the largest known crater 130 miles across, although there are hints along the edge of the daylight side of a crater possibly as much as 240 miles across. Only 20% of the surface has been mapped. Like Ariel, the trailing hemisphere shows a concentration of CO_2 ice, with water ice on the leading hemisphere.

Umbriel was a melancholy spirit in Pope.

[U 17] Titania

William Herschel discovered Uranus III nearly six years after discovering the planet, on January 11, 1787. Titania is 979 miles in diameter, the largest of Uranus's moons. The orbit's semimajor axis is 270,700 miles, taking 8.7058726 days. The inclination is 0.34 degrees. An albedo of 17 gives it an apparent magnitude of 13.5.

Titania, U 17, the largest moon of Uranus, from 300,000 miles

Titania has craters as much as 200 miles in diameter. There are scarps and canyons, including one canyon that runs for over 900 miles from the equator to the south pole. Titania is believed to be about half water ice and half rocky material, with an interface between the rocky core and ice mantle of about 30 miles, which could be liquid water. The surface shows fewer craters than either Umbriel or Oberon, suggesting a younger surface. Voyager 2's closest approach while photographing Titania was 226,900 miles.

Titania was queen of the fairies in "Midsummer Night's Dream".

[U 18] Oberon

Oberon, Uranus IV, was discovered by William Herschel the same day he discovered Titania. It is 946 miles in diameter. It is 362,369 miles from Uranus, taking 13.46924 days to orbit. The inclination is 0.058 degree. The albedo is 14, with a density of 1.63, and an apparent magnitude of 13.7. It has the reddest tone of Uranus's moons. The surface gravity is 3% that of Earth.

Oberon has craters up to 130 miles in diameter, and there is some evidence that the unseen portion may have a crater as much as 230 miles across with a central peak seven miles high. The surface seems to have water ice as well as rocks and some carbon compounds.

Oberon was King of the fairies and Titania's consort in "Midsummer Night's Dream".

[U 19] Francisco

Uranus XXII was given the name Francisco after its August 13, 2001 discovery by Matthew J. Holman, Brett Gladman, and John Kavelaars. Its orbit ranges from 2,268,000 to 8,042,800 miles from Uranus, with a period of 266.56 days. The inclination is 147.459 degrees, making it retrograde with respect to the planet and inner moons, but direct for the rest of the Solar System. The diameter is thirteen miles. It is the first moon of Uranus not in synchronous rotation.

Francisco is named for a lord in "The Tempest".

[U 20] Caliban

Brett Gladman, John Kavelaars, Philip Nicholson, and Joseph Burns discovered Uranus XVI on September 6, 1997, using the 200 inch telescope at Mount Palomar. It is 45 miles in diameter. The orbit of Caliban ranges from 3,777,816 to 5,203,086 miles, taking 579.73 days to go around the planet. The inclination is 120.28 degrees. It rotates in just 2.7 hours, the shortest known period for any moon.

Caliban is named for the monstrous servant of Prospero in "The Tempest".

[U 21] Stephano

Stephano is Uranus XX. It was discovered by Brett Gladman, John Kavelaars, Matthew Holman, Jean Marc Petit and Hans Scholl on July 18, 1999. Stephano has a 20 mile diameter. It takes 677.37 days for an orbit ranging from 4,181,663 to 5,789,290 miles from the planet. The inclination is 144 degrees.

Stephano was Prospero's drunken butler in "The Tempest".

[U 22] Trinculo

Trinculo, or Uranus XXI, was discovered by Matthew Holman, John Kavelaars, Brett Gladman, and Dan Milisavljevic on August 13, 2004. The orbit of 3,662,825 to 6,899,153 miles takes 749.24 days (2 years 19.24 days, Earth time). The inclination is 164 degrees.

Trinculo was the drunken jester in "The Tempest".

[U 23] Sycorax

Sycorax, the largest of Uranus's distant moons at 95 miles diameter, was discovered September 6, 1997. A team of Philip Nicholson, Brett Gladman, Joseph Burns, and John Kavelaars used the 200 inch telescope at Mount Palomar. Sycorax runs from 2,612,165 to 11,514,153 miles from the planet in a 1288.28 day orbit. The inclination is 152.5 degrees. The apparent magnitude fluctuates by 0.07 around a mean of +20.8 in 3.6 hours, suggesting that this is the rotational period. The surface has a light red tint. It is also Uranus XVII.

Sycorax is mentioned in "The Tempest" as Caliban's (U 20) mother, but does not appear as a character.

[U 24] Margaret

Scott Sheppard and David Jewitt found this twelve mile diameter moon, Uranus XXIII, on August 29, 2003. It is the only distant moon of Uranus not in a retrograde orbit. It goes from 3,021,677 to 14,794,813 miles, but these figures are only means, as the orbit is profoundly affected by the gravitational influence of the Sun, and both periapsis and apapsis can change from one orbit to the next by many thousands of miles. The orbital period averages 1687 days (4 years 226 days). The inclination is 51 degrees.

Margaret was a servant in "Much Ado About Nothing".

[U 25] Prospero

Prospero, or Uranus XVIII, was discovered on July 18, 1999 by Matthew Holman, John Kavelaars, Brett Gladman, Jean-Marc Petit, and Hans Scholl. It is 31 miles in diameter, and has a neutral gray color. The orbit ranges from 5,604,731 to 14,585,221 miles in 1978.3 days. The inclination is 146.07 degrees.

Prospero was the main character in "The Tempest".

[U 26] Setebos

John Kavelaars, Brett Gladman, Matthew Holman, Jean-Marc Petit and Hans Scholl discovered Uranus XIX of July 18, 1999. This moon is 30 miles in diameter and neutral gray in color. It ranges from 4,419,700 to 17,213,500 miles from the planet in an orbit taking 2225 days (6 years 34 days). The inclination is 146 degrees.

Setebos was the name of the god worshipped by Caliban (U 20) and his mother Sycorax (U 23) in "The Tempest".

[U 27] Ferdinand

Ferdinand, or Uranus XXIV, was discovered August 13, 2001 by Matthew Holman, John Kavelaars, Brett Gladman, and Dan Milisavljevic. It has a twelve mile diameter. The orbit ranges from 8,200,000 to 17,758,200 miles. This takes 2805 days (7 years 249 days). It has a neutral gray color.

Ferdinand was a son of the King of Naples in "The Tempest".

NEPTUNE

[N 1] Naiad

Naiad was discovered in September 1989, the third known moon, and so Neptune III. It is 59 by 37 by 32 miles. It orbits the planet at 29,950 miles, with a low eccentricity varying that by only six miles either way. This takes 0.249 day to circle Neptune with an inclination of 4.75 degrees. Rotation and revolution periods are the same (synchronous) with the long axis pointed to the planet. The apparent magnitude at opposition is 23.9.

In Greek mythology naiads were nymphs controlling streams, brooks, springs, wells, and fountains, all forms of fresh running water, and patrons of music and poetry.

[N 2] Thalassa

Richard Terrile found Thalassa, Neptune IV, on Voyager 2 photographs in September 1989. It is 66 by 62 by 32 miles, and orbits 31,100 miles from Neptune. The 0.21 degree inclined orbit takes 0.3115 days. The albedo is 9 and the apparent magnitude 23.3. Like all small close moons, it rotates synchronously with the long axis permanently pointed to the planet.

Thalassa was the Ancient Greek word for ocean, and was also the name of a sea goddess.

[N 3] Despina

Neptune V was found on Voyager 2 photographs in July 1989. It is 112 by 92 by 78 miles. The inclination is 0.216 degree. Despina has an albedo of 9, and an apparent magnitude of 22. The orbit's semimajor axis is 32,620 miles with an eccentricity so slight that it varies by only four miles either way. The period is 0.335 day.

Despina was a sea nymph and daughter of Poseidon.

[N 4] Galatea

Stephen Synnott found Galatea, or Neptune VI, on Voyager 2 photographs in July 1989. It is 127 by 114 by 89 miles. Its orbit of 38,473 miles varies by only two miles either way. The orbit takes 0.419 day to go around the planet, with an inclination of just 0.052 degree to Neptune's equator. Galatea is a shepherd moon for a ring 620 miles further from Neptune. This moon has the very low density of 0.75, strongly suggesting that it is a rubble heap with large vacant spaces within it. This density is actually lower than

Saturn's, a planet which is all gases. The surface gravity would be 0.2% of Earth's. The albedo is 8, and apparent magnitude is 23.8.

Galatea was a sea nymph and the lover of the cyclops Polyphemus, who had an unfortunate run-in with Ulysses.

[N 5] Larissa

Larissa, or Neptune VII, was discovered May 29, 1981 by Harold Reitsema, William Hubbard, Larry Lebofsky and David Tholin. Larissa is 134 by 127 by 104 miles. Its orbit's semimajor axis is 45,673 miles with periapsis and apapsis varying from that by just 80 miles either way. The orbit takes 0.555 day. Inclination is 0.251 degree. Larissa is heavily cratered. The apparent magnitude is 21.5.

Larissa was a nymph and lover of Poseidon.

[N 6] Proteus

Proteus is Neptune VIII, and was discovered on Voyager 2 photographs by Stephen Synnott on June 16, 1989. It is 271 by 258 by 250 miles, second largest of Neptune's moons. However, it is not easily seen in telescopes from Earth because the much brighter planet is so close. It is 73,058 plus or minus 39 miles from Neptune. The orbital period is 1.1223 days. The orbital inclination is 0.524 degree.

Proteus has many craters, the largest known being more than 130 miles across and eight miles deep. This moon is dark gray with an albedo of 9.6, and magnitude 19.7. The spectrum suggests a surface of hydrocarbons, and no water ice has been detected. There are grooves and valleys, some related to large craters, and one of which runs parallel to the equator.

Proteus was the shape-shifting sea god with some oracular powers. He is usually depicted as an old man concerned for the welfare of sea creatures.

[N 7] Triton

Triton was discovered by William Lassell on October 10, 1846, less than two weeks after Neptune itself was discovered, and so is also Neptune I. It has a diameter of 1680 miles, one of the larger moons of the Solar System. It has a retrograde orbit 220,305 miles from Neptune, taking 5.877 days (5 days 21 hours 2 minutes 53 seconds) to go around Neptune in the reverse direction from that of other moons that are near their planets. The inclination is therefore greater than ninety degrees, at 156.885 degrees. Triton's unusual orbit is believed to be a consequence of the moon having originally been an independent object that was captured by Neptune early in the history of the Solar System.

Triton has a high albedo of 76, and an apparent magnitude of 13.4. The surface is mostly frozen nitrogen, N2 ice, overlying a crust of water ice, but methane and carbon dioxide have also been detected. All are frozen solid. The core is rock and metal. Triton has a density of 2.061, and is thought to be about 30% water ice. When originally captured by Neptune the orbit, now nearly perfectly circular, would have been extremely elliptical, leading to internal heating with consequent differentiation, so denser materials settled to the core. The denser materials would presumably include radioactives, which would also provide some heating. Radiogenic heating would be effective equally around Triton, but tidal heating would be more effective near the poles, making the crust thinner in polar regions. Heating may also explain why Triton shows few large craters, and is relatively flat, with no point rising much higher or lower than half a mile. It is likely no point on the surface is older than 50 million years. The surface has a reddish tint caused by solar ultraviolet light converting methane ice (CH_4) to tholins.

The dark spots are some of the spouts from Triton's interior

The atmosphere is nearly all N2, but very thin, about 7 X 10^-4 Earth's normal at sea level. The surface gravity is 0.08G, and the escape velocity is 0.904 miles per second.

Voyager 2 spotted cryovolcanoes and geysers on Triton. Geyser plumes rose to a height of five miles, and dust streaks were seen on the surface to extend 90 miles downwind from the geysers and volcanoes. The geysers mostly spew N2. The volcanic lavas are a mix of water and ammonia (H2O + NH3). Triton could have a layer of liquid water with up to 15% ammonia. Such mixes lower the temperature at which water freezes, so the assumed liquid water layer could exist at a temperature as cold as –140F.

In mythology all the sons of Neptune were called tritons, and are sometimes depicted as riding on dolphins.

[N 8] Nereid

Nereid (neer ee id) is Neptune II. It was discovered by Gerard P. Kuiper (the astronomer for whom the Kuiper Belt is named) on May 1, 1949. It has a diameter of 211 miles, third largest of Neptune's moons. It takes 360.136 days for an extremely eccentric orbit of 853,000 to 5,998,750 miles. The inclination is 32.55 degrees, the only distant moon not in a retrograde orbit. This orbit is believed to have resulted from interactions when Triton was captured, and it is assumed other moons were lost at the same time, either colliding to form some of Neptune's rings, escaping to become Kuiper Belt asteroids, or falling into the planet. Nereid has a surface of water ice, an albedo of 15.5, magnitude 19.2, and may rotate in 13.6 hours. The Voyager spacecraft never got closer than 2.9 million miles, limiting the quality of information obtained.

Nereids were daughters of the sea god Nereus. There were supposed to be fifty of them, but various ancient writers provided nearly one hundred different names.

[N 9] Halimede

Halimede (hal ih mee dee) was discovered August 14, 2002 by Matthew Holman, John Kavelaars, T. Grav, William Fraser, and Dan Milisavljevic. It has a 38 mile diameter, and a color similar to that of Nereid. The orbit ranges from 7,585,000 to 13,045,000 miles from the planet. This takes 1879 days (5 years 53 days). The inclination is 112.7 degrees. It is Neptune IX.

Halimede was one of the fifty Nereids.

[N 10] Sao

Sao was discovered August 14, 2002 at the same time and by the same people who discovered Halimede, making it Neptune XI. It is slightly the smaller of the two with a diameter of 27 miles. Sao orbits Neptune from 11,919,000 to 15,687,800 miles in a period of 2912.72 days (7 years 357.7 days). The inclination is 53.483 degrees.

Sao was the Nereid of sailing ships, responsible for the safety of sailors. However, SAO is also the abbreviation for the Smithsonian Astrophysical Observatory, so it is likely this helped the choice of name.

[N 11] Laomedeia

Laomedeia was discovered August 13, 2002 by Holman, Kavelaars, Grav, Fraser, and Milisavljevic. It has a diameter of 26 miles and orbits Neptune from 8,826,400 to 20,443,800 miles. This takes 3171.33 days. The inclination is 37.874 degrees. It is also Neptune XII.

Laomedeia was another of the Nereids.

[N 12] Psamathe

Psamathe (sa math ee) was discovered in 2003 by Scott Shepherd, David Jewitt, and Jan Klayna. The diameter is 24 miles. The orbit ranges from 18,491,000 to 41,294,200 miles, and takes 24 years 310.7 days (9074.7 days) to complete. The inclination is 126.3 degrees. It is Neptune X. This is the second largest orbit of any moon.

Psamathe was a Nereid.

[N 13] Neso

Neso was discovered in 2002 by Matthew Holman, John Kavelaars, Brett Gladman, T. Grav, W. Fraser, and Dan Milisavljevic. It has a diameter of 38 miles. The orbit from 13,188,000 to 47,094,000 miles is the largest orbit of any moon in the Solar System. Even Jupiter could not hold onto a moon 47 million miles away, but the Sun's gravity is a lot weaker at Neptune's distance, which is seven times further from the Sun than is Jupiter. Neso takes 26 years 250 days (9740.7 days) for a complete orbit. The inclination is 136.439 degrees.

Neso was a Nereid.

MOONS OF DWARF PLANETS

PLUTO

[P 1] Charon

Charon was discovered on June 22, 1978 by James Christy, working at Flagstaff, Arizona. Charon is 12,160 miles from Pluto, but only 10,290 miles from the barycenter of the Pluro-Charon binary system. On this basis, this is the only known barycenter in the Solar System to lie outside a primary body in open space. Charon's orbit requires 6.387 days (6 days 9 hours 17 minutes 36.7 seconds). The orbital inclination is 0.001 degree. The density is 1.65, with an albedo of 38. It is believed Charon is 55% rock and 45% ices. The surface appears to be water ice and ammonia hydrates. There may be cryovolcanoes similar to those found on Triton (Neptune). The apparent magnitude varies from 16.8 to 17.3 because of Pluto's extremely elliptical orbit, and the surface gravity is 0.028 G. The equator appears brighter than the polar caps, with the south pole especially dark.

Charon was the boatman who carried souls across the River Styx to the underworld ruled by Pluto. The name is properly pronounced karen, but the discoverer prefers sharon to resemble his wife's name, Charlene.

[P 3] S/2012 P-1

The discovery of this moon was announced July 10, 2012 by Mark Showalter, S. Alan Stern, Marc Bluie, Andrew Steffl, and Max Mutchler, based on observations using the Hubble Space Telescope from June 29 to July 9. At magnitude 27, it has a diameter between 6 and 15 miles. The orbit is roughly 26,080 miles from Pluto, taking 20.2 days to go around the planet. Cerberus won a poll of the public as a possible name.

[P 2] Nix

Nix was discovered June 14, 2005 by Hal Weaver, S. Allen Stern, Max Mutchler, Andrew Steffl, Marc Bluie, William Merline, John Spencer, Eliot Young, and Leslie Young. Nix is 30,250 miles from Pluto, plus or minus twenty miles. It takes 24.856 days to orbit the barycenter, with an inclination of 0.195 degree. The apparent magnitude is 23.5. Little has been determined about the size of Nix, although estimates suggest a diameter in the neighborhood of 30 to 40 miles.

Nyx was the mother of Charon, Hypnos (sleep) and Thanatos (death). The alternate spelling was chosen to avoid confusion with an asteroid named Nyx.

[P 4] S/2011 P-1

S/2011 P-1 was discovered on June 28, 2011 by Mark Showalter. Its diameter is estimated at around 18 miles. It is 36,600 miles from the barycenter with an orbit that takes 32.1 days with an inclination close to zero. The apparent magnitude is 26.1

Showalter has said he will propose an appropriately Hades related name. He also has suggested there are hints of two more very small moons for Pluto (this could presumably include the moon listed here as P 3). William Shatner influenced a public poll to try to get this moon named Vulcan, a nephew of Pluto.

[P 5] Hydra

Hydra was discovered in 2005 at the same time as Nix. It is believed to be the larger of the two, with an estimated diameter of 60 miles. The orbit is 40,209 miles (plus or minus 100) taking 38.206 days. The inclination is 0.212 degree. The apparent magnitude is 23.3.

Hydra was the serpent that wrestled with Hercules. It has nothing mythological to do with Pluto, but the name was chosen, with that of Nix, to honor with the moons' initials the *New Horizons* spacecraft on its way to Pluto.

HAUMEA

[H 1] Namaka

Namaka was discovered June 30, 2005 by Michael Brown, Chad Trujillo, and David Rabinowitz. It orbits 15,933 miles from Haumea, plus or minus 400 miles. The orbital period is 18.2783 days, with an inclination of 113.013 degrees. The surface is mainly water ice.

Namaka was a daughter of Haumea, goddess of the sea and of birth for the native Hawaiians.

[H 2] Hi'iaka

Hi'iaka was discovered January 25, 2005 by Michael Brown, Chad Trujillo and David Rabinowitz. It is 30,975 miles plus or minus 120 miles from Haumea. The diameter is about 217 miles. The orbital period is 49.12 days, with an inclination of 126.356 degrees. Density is about 1.0. The surface is water ice. Apparent magnitude is 20.6.

Hi'iaka was the patron goddess of the Big Island of Hawaii.

ORCUS

[O 1] Vanth

Vanth was discovered November 13, 2005 by Michael Brown and T.-A. Suer. It is 5600 miles from Orcus, plus or minus 40 miles. The orbital period is 9.54 days. The inclination is 21 degrees. Rotation is synchronous. Vanth has an albedo of 12, and a reddish surface quite unlike the color of Orcus. This takes it out of the realm of moons that were part of their primary but knocked loose, and makes it most likely a captured object. The apparent magnitude is 22.

Vanth was an Etruscan goddess who guided souls of the dead to the underworld.

Note: The IAU does not recognize Orcus as a dwarf planet yet, but it seems to fulfill all the requirements, so in anticipation of an upgrade, this entry is included.

QUAOAR

[Q 1] Weywot

Weywot was discovered February 22, 2007 by Michael Brown. It is 9000 miles, plus or minus 900 miles, from Quaoar. The orbital period is 12.438 days, with an inclination of 14 degrees. Weywot has a diameter estimated at around 46 miles. The apparent magnitude is 24.9.

Weywot was a son of the sky god Quaoar.

ERIS

[E 1] Dysnomia

Dysnomia was discovered September 10, 2005 by Antonine Bouchez, Michael Brown, Chad Trujillo and David Rabinowitz. It is 23,200 miles from Eris, with an inclination of 142 degrees. This orbit takes 15.774 days. Dysnomia has a diameter estimated at 200 miles, and an apparent magnitude of 23.1.

Dysnomia was a daughter of Eris and as much a trouble maker as her mother. Brown prefers the long eye pronunciation of the Y in the name, suggesting his wife Diane.

ASTEROIDS

As of this writing about 200 asteroids have been found to have one or more moons. Most seem to be pieces knocked off the asteroid during a collision. Asteroidal moons will be dealt with in a book on asteroids.

Alphabetical List of Named Moons

A

Adrastea	J 2		Caliban	U 20
Aegaeon	S 9		Callirrhoe	J 47
Aegir	S 52		Callisto	J 8
Aitne	J 34		Calypso	S 17
Albiorix	S 30		Carme	J 46
Amalthea	J] 3		Carpo	J 14
Ananke	J 30		Chaldene	J 38
Anthe	S 12		Charon	P 1
Aoede	J 43		Cordelia	U 1
Arche	J 58		Cressida	U 4
Ariel	U 16		Cupid	U 9
Atlas	S 4		Cyllene	J 53
Autonoe	J 64			

B

Bebhionn	S 32
Belinda	U 10
Bergelmir	S 45
Bestla	S 54
Bianca	U 3

C

D

Daphnis	S 3
Deimos	M 2
Desdemona	U 5
Despina	N 3
Dione	S 18
Dysnomia	E 1

E

Elara	J 13
Enceladus	S 14
Epimetheus	S 7
Erinome	J 42
Erriapus	S 33
Euanthe	J 22
Euclade	J 54
Euporie	J 16
Europa	J 6
Eurynome	J 48

F

Farbauti	S 50
Fenrir	S 57
Ferdinand	U 27
Fornjot	S 62
Francisco	U 19

G

Galatea	N 4
Ganymede	J 7
Greip	S 38

H

Halimede	N 9
Harpalyke	J 31
Hati	S 48
Hegemone	J 57
Helene	S 19
Heliki	J 23
Hermippi	J 28
Herse	J 33
Hi'iaka	H 2
Himalia	J 11
Hydra	P 5
Hyperion	S 23
Hyrrokin	S 39

I

Iapetus	S 24
Ijiraq	S 26
Io	J 5
Iocaste	J 25
Isonoe	J 59

J

Janus	S 8
Japetus	*see Iapetus*

J (continued)

Jarnsaxa	S 40
Jocasta	*see Iocaste*
Juliet	U 6

K

Kale	J 35
Kaliki	J 45
Kallichore	J 44
Kari	S 59
Kiviuq	S 25
Kore	J 52

L

Larissa	N 5
Leda	J 10
Loge	S 61
Lysithia	J 12

M

Mab	U 13
Margaret	U 24
Megaclite	J 65
Methone	S 11
Metis	J 1
Mimas	S 10
Miranda	U 15
Mneme	J 29
Mundilfari	S 42

N

Naiad	N 1
Namaka	H 1
Narvi	S 46
Nereid	N 8
Nix	P 2

O

Oberon	U 19
Ophelia	U 2
Orthosie	J 24

P

Paaliaq	S 28
Pallene	S 13
Pan	S 2
Pandora	S 6
Pasiphae	J 56
Pasithee	J 50
Perdita	U 11
Phobos	M 1

Phoebe	S 27
Polydeuces	S 20
Portia	U 7
Praxidiki	J 27
Prometheus	S 5
Prospero	U 25
Proteus	N 6
Puck	U 12

Q

R
Rhea	S 21
Rosalind	U 8

S
Setebos	U 26
Siarnaq	S 35
Sinope	J 62
Skathi	S 29
Skoll	S 34
Sponde	J 63
Stephano	U 21
Surtur	S 58
Suttugr	S 47
Sycorax	U 23

T
Tarqeq	S 36
Tarvos	S 41
Taygete	J 36
Telesto	S 16
Tethys	S 15
Thalassa	N 2
Thebe	J 4
Thelxinoe	J 21
Themisto	J 9
Thrymer	S 51
Thyone	J 32
Titan	S 22
Titania	U 18
Trinculo	U 22
Triton	N 7

U
Umbriel	U 17

V
Vanth	O 1

W
Weywot	Q 1

X

Y
Ymir	S 60

Z

UNNAMED
S/2000	J-11	J 13.5
S/2003	J-2	J 66
S/2003	J-3	J 12
S/2003	J-4	J 55
S/2003	J-5	J 61
S/2003	J-9	J 60
S/2003	J-10	J 40
S/2003	J-12	J 15
S/2003	J-15	J 39
S/2003	J-16	J 26
S/2003	J-18	J 18
S/2003	J-19	J 37
S/2003	J-23	J 41
S/2004	S-7	S 55
S/2004	S-12	S 49
S/2004	S-13	S 37
S/2004	S-17	S 44
S/2006	S-3	S 56
S/2006	S-1	S 43
S/2007	S-2	S 31
S/2007	S-3	S 53
S/2009	S-1	S 1
S/2010	J-1	J 49
S/2010	J-2	J 20
S/2011	J-1	J 19
S/2011	J-2	J 51
S/2011	P-1	P 3
S/2012	P-1	P 5

GROUPS
Albiorix	S 30, 33, 41
Alkyonides	S 11, 12, 13

Ananke	J 16, 17, 18, 20 to 32, 39, 41
Carme	J 33 to 38, 40, 42, 44, 45, 46, 49, 50, 54, 58 to 61
Galilean	J 5, 6, 7, 8
Himalia	J 10, 11, 12, 13, 13.5
Kiviuq	S 25, 28, 35
Pasiphae	J 43, 47, 48, 52, 53, 55, 56, 57, 61 to 66

SHEPHERDS and RING SOURCES

J2
S2, 3, 5, 6, 9, 10
U 1, 2
N 1, 2, 3, 4

CO-ORBITALS AND TROJANS

S 7, 8; 15, 16, 17; 18, 19, 20

MOONS SHARING RESONANCES

J 5, 6, 7
S 5, 6; 6, 10; 9, 10; 10, 15; 22, 23
U 10, 11

MOONS WITH DIAMETERS OVER 1000 MILES

T 1
J 5, 6, 7, 8
S 22
N 7

MOONS WITH DIAMETERS FROM 250 MILES TO 1000 MILES

Moons smaller than this generally have an irregular shape.

J 3
S 10, 14, 15, 18, 21, 24, 27
U 14, 15, 16, 17, 18
N 6, 8
P 1

MOONS WITH SYNCHRONOUS ROTATION

(that means they rotate and revolve in the same time, keeping one face always to their planet)

T 1
M 1, 2
J 1 through 8

S 1 through 22, 24
U 1 through 18
N 1 through 7
P 1 (2,3,4,5 are possible but unknown)
H 1, 2?
O 1
E 1

MOONS WITH RETROGRADE ORBITS

J 15 through 67
S 27, 29, 31, 34, 37 to 40, 42 to 62
U 19, 20, 21, 22, 23, 25, 26
N 7, 9, 12, 13
H 1, 2
E 1

MOONS BRIGHTER THAN MAGNITUDE +9.5 AT OPPOSITION

T 1
J 5, 6, 7, 8
S 22

CRYOVULCANISM AND GRAVITY PUMPING-INDUCED LIQUIDS

J 5, 6? ,7?
S 14, 22?
N 7
P 1

NAMED SURFACE FEATURES MENTIONED

Aitkin South Pole Basin	T1
Asgard Basin	J 8
Bogle Crater	U 12
Cassini Regio	S 24
Evander Crater	S 18
Herschel Crater	S 10
Mare Nubium	T 1
Mare Orientale	T 1
Mare Tranquillitatis	T 1
Oceanus Procellarum	T 1
Odysseus Crater	S 15
Surtur volcano	J 5 (mention is at S 58)
Stickney Crater	M 1
Swift Crater	M 2
Valhalla Basin	J 8
Voltaire Crater	M 2

Zethus   J 4

DISCOVERERS AND OTHER PEOPLE MENTIONED

A
Alexandersen, Mike   J 49
Asimov, Isaac   Mercury

B
Barnard, Edward   E. J 3
Bluie, Marc W.   P 3
Bond, George   S 23
Bond, William   S 23
Bouchez, Antonine   E 1
Brown, Michael   H 1, 2; O 1; Q 1
Brozovic, Marina   J 13.5, 49
Burns, Joseph   U 20, 23

C
Cassini, Giovanni   S 18, 21, 24
Christy, James   P 1
Collins   S 5

D
deCamp, L. Sprague   S 58
Dollfus, Audouin   S 8

E

F
Fernandez, Y. R.   J 65
Flammarion, Camille   J 3
Fountain, John   S 7
Fraser, William   N 9, 10, 11, 13

G
Galilei, Galileo   T 1; J 5, 6, 7, 8
Gehrels, Tom   J 47
Gladman, Brett   J 18, 26, 33, 37; S 25, 26, 28, 29, 33, 35, 41,
                          42, 47, 49, 51, 60; U 18 to 23, 25 to 27; N 13
Grav, T.   N 9, 10, 11, 13

H
Hall, Angelina S.   M 1, 2
Hall, Asaph   M 1, 2
Heinlein, Robert A.   J 7, 10
Herschel, John   S 15
Herschel, William   S 10, 14; U 17, 18
Holman, Matthew   S 30; U 19, 21, 22, 25 to 27; N 9, 10, 11, 13
Hubbard, William   N 5
Huyghens, Christiaan   S 22

J

Jacobson, Robert	J 13.5, 49
Jewitt, David	J 2, 64, 65; S 31, 32, 36 to 40, 43 to 46, 48 to 50, 52 to 59, 61, 62; U 24; N 12

K

Karkoschka, Erich	U 11
Kavelaars, John	S 25, 26, 28, 29, 33, 35, 41, 42, 47, 51, 60; U 19 to 23, 25 to 27; N 9, 10, 11, 13
Kayna, Jan	J 64; S 31, 32, 36 to 40, 43 to 46, 48 to 50, 52 to 59, 61, 62; N 12
Kepler, Johannes	M 1, 2
Kowal, Charles	J 9, 10
Kuiper, Gerard	U 14; N 8

L

Lacacheux, Jean	S 19
Lagrange, Joseph-Louis	S 16
Laques, Pierre	S 19
Larson, Stephen	S 7
Larson, T. A.	J 47
Lassell, William	S 23; U 15, 16; N 7
Lebofsky, Larry	N 5
Lissauer, Jack	U 7, 13

M

Maguier, C.	J 65
Marsden, Brian	S 31, 32, 34, 36 to 40, 43 to 46, 48 to 50, 52 to 59, 61, 62
McMillan, R. S.	J 47
Melotte, Philibert	J 56
Merline, William	P 2
Milisavljevic, Dan	U 22, 27; N 9, 10, 11, 13
Montani, J.	J 47
Morabito, Linda	J 5
Mutchler, Max	P 2

N

Nicholson, Philip	U 20, 23
Nicholson, Seth	J 12, 30, 46, 46

O

P

Perrine, Charles	J 11, 13
Pessau, Dan	S 17
Petit, Jean Marc	U 21, 25, 26
Pickering, William	S 27
Porco, Carolyn	S 1, 3, 9
Pratt, Fletcher	S 58

CPSIA information can be obtained
at www.ICGtesting.com
Printed in the USA
BVHW060515140219

540202BV00006B/638/P